SpringerBriefs in Computer Science

More information about this series at http://www.springer.com/series/10028

Rosario Aragues · Carlos Sagues
Youcef Mezouar

Parallel and Distributed Map Merging and Localization

Algorithms, Tools and Strategies
for Robotic Networks

 Springer

Rosario Aragues
Instituto de Investigación en Ingeniería de
  Aragón
University of Zaragoza
Saragossa
Spain

Carlos Sagues
Instituto de Investigación en Ingeniería de
  Aragón
University of Zaragoza
Saragossa
Spain

Youcef Mezouar
Institut Pascal, CNRS
Clermont Université, IFMA
Clermont-Ferrand
France

ISSN 2191-5768   ISSN 2191-5776   (electronic)
SpringerBriefs in Computer Science
ISBN 978-3-319-25884-3   ISBN 978-3-319-25886-7   (eBook)
DOI 10.1007/978-3-319-25886-7

Library of Congress Control Number: 2015952999

Springer Cham Heidelberg New York Dordrecht London

Springer International Publishing AG Switzerland is part of Springer Science+Business Media
(www.springer.com)

Preface

The increasing interest in multi-robot systems is motivated by the wealth of possibilities offered by teams of robots cooperatively performing collective tasks. In these scenarios, distributed strategies attract a high attention, especially in applications which are inherently distributed in space, time, or functionality. These distributed schemes not only reduce the completion time of the task due to the parallel operation, but also present a natural robustness to failures due to the redundancy. In addition to the classical issues associated to the operation of individual robots, these scenarios introduce novel challenges specific to communications and coordination of the members of the robot team.

In this book, we analyze a particular problem of high interest in these scenarios: distributed map merging and localization. It allows the robots to acquire the knowledge of their surrounding needed for carrying out other coordinated tasks. We identify the main issues associated to this problem, and we present at each chapter different distributed strategies for solving them.

The explanation of this problem serves us as a tool for discussing topics which are classical in these scenarios and for introducing the reader to several multi-robot concepts. Thus, this book has several purposes. First, to give a complete solution to the distributed map merging and localization problem, which can be implemented in a multi-robot platform. Second, to provide the reader with the necessary tools for proposing new solutions to the multi-robot perception problem, or for addressing other interesting topics related to multi-robot scenarios. And third, to attract the attention to multi-robot systems and distributed strategies.

The authors have been working in different topics related to robotics perception and control. In this book they analyze distributed algorithms for perception in localization and map merging. The authors believe that this is an interesting topic, and that there are still many challenges that remain to be addressed in order to achieve the final aim of having a complete availability of these systems in the life of human beings.

This book can be of interest to the robotics and control communities, to post-graduate students and researchers, and, in general, to anyone interested in multi-robot systems. We do not make any assumption about the background needed to read the book. However, the basic understanding on mathematics of a graduate student is necessary. It is very difficult to give a fully self-contained material and, although we have introduced as many explanations and demonstrations as we could, we give references which can be studied if needed.

Saragossa, Spain   Rosario Aragues
Saragossa, Spain   Carlos Sagues
Clermont-Ferrand, France   Youcef Mezouar
October 2013

Contents

Chapter 1
Introduction

Abstract The increasing interest in multi-robot systems is motivated by the wealth of possibilities offered by teams of robots cooperatively performing collective tasks. This chapter introduces the multi-robot map merging and localization problem, and makes a revision of the state of the art in the topics involved. The last section in this chapter contains the book organization and explains the way in which the authors have focused the study.

Keywords Networked robots · Distributed systems · Parallel computation · Limited communication · Multi-robot perception · Localization · Data association · Map merging

1.1 Motivation

The increasing interest in multi-robot applications is motivated by the wealth of possibilities offered by teams of robots cooperatively performing collective tasks. The efficiency and robustness of these teams go beyond what individual robots can do. In these scenarios, distributed strategies attract a high attention, especially in applications which are inherently distributed in space, time or functionality. These distributed schemes do not only reduce the completion time of the task due to the parallel operation, but also present a natural robustness to failures due to the redundancy. In addition to the classical issues associated to the operation of individual robots, these scenarios introduce novel challenges specific to communications and coordination of the members of the robot team. Several distributed algorithms are based on behaviors observed in nature. It has been observed that certain groups of animals are capable of deploying over a given region, assuming a specified pattern, achieving rendezvous at a common point, or jointly initiating motion or changing direction in a synchronized way (Fig. 1.1).[1] Species achieve synchronized behavior, with limited

---

[1]The images in Fig. 1.1 have been obtained from the following sources. Figure 1.1 (left): https://commons.wikimedia.org/wiki/File:Grus_grus_flocks.jpg; Andreas Trepte, http://www. photo-natur.de. Figure 1.1 (right): https://commons.wikimedia.org/wiki/File:IRobot_Create_team. jpg; Jiuguang Wang; A team of iRobot Create robots at the Georgia Institute of Technology.

© The Author(s) 2015
R. Aragues et al., *Parallel and Distributed Map Merging and Localization*,
SpringerBriefs in Computer Science, DOI 10.1007/978-3-319-25886-7_1

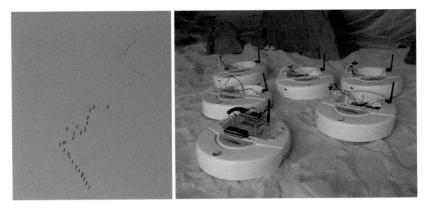

Fig. 1.1 Examples of pattern formation observed in animals and of multi-robot teams

sensing or communication between individuals, and without apparently following the instructions of a group leader. Robotic researchers have intensively investigated on coordination strategies for multi-robot systems (Fig. 1.1) capable of imitating these collective behaviors. In particular, it is worth mentioning the following strategies: rendezvous, which consists of the robots getting together at a certain location; deployment or coverage, which consists of deploying the robot team over the region of interest, and agreement, which consists of reaching consensus upon the value of some variable. Agreement has a special interest and recently it has been shown that several multi-robot strategies, including pattern formation and rendezvous, can be transformed into an agreement problem.

Our research is focused on distributed applications for perception tasks. Perception is of high importance in robotics, since almost all robotic applications require the robot team to interact with the environment. Then, if a robot is not able to obtain an environmental representation from others, or an a priori representation is not available, it must have perception capabilities to sense its surroundings. Perception has been long studied for single robot systems and a lot of research has been carried out in the fields of localization, map building, and exploration. Among the different sensors that can be used to perceive the environment, visual perception using conventional or omnidirectional cameras has been broadly used because of its interesting properties (Fig. 1.2).

While the first kind of cameras (Fig. 1.3) are widely known and used in any area, omnidirectional devices are very popular in robotic applications. These cameras are able to capture visual information within 360° around the robot due to the use of an hyperbolic mirror (Fig. 1.4). Cameras provide bearing-only information through the projection of landmarks which are in the scene. In order to recover the position of these landmarks in the world, multiple observations taken from different positions must be combined. The manipulation of bearing data is an important issue in robotics. Compared with information extracted from other sensors, such as lasers, bearing information is complicated to use. However, the multiple benefits of using cameras

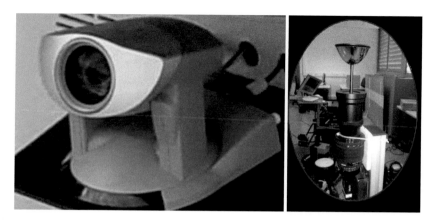

Fig. 1.2   Examples of conventional (*left*) and omnidirectional cameras (*right*)

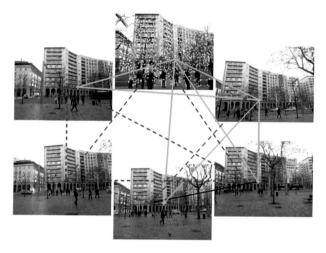

Fig. 1.3   Examples of images taken by a team of six robots moving in formation equipped with a conventional camera. Crosses are features extracted from the images and lines between images represent features matches

Fig. 1.4   Examples of omnidirectional images. Crosses are features extracted form the images, and lines between images represent features matches

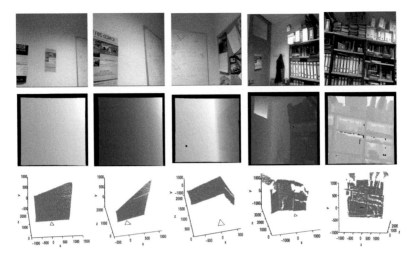

Fig. 1.5 An example of the images obtained with the RGB-D sensor

have motivated the interest in the researchers. These benefits include the property that cameras are able to sense quite distant features, so that the sensing is not restricted to a limited range. An additional kind of cameras of high interest are RGB-D devices. They provide both regular RGB (Fig. 1.5, first row) and depth image information (Fig. 1.5, second row). Thus, it is possible to compute the landmark 3D position from a single image (Fig. 1.5, third row).

Robots sense the environment and combine the bearing data to build representations of their surroundings in the form of stochastic maps. Each individual robot perceives the portion of the environment where it is operating. In order to make decisions in a coordinated way, the robots must merge their local observations into a global map. We can distinguish between centralized and distributed approaches. Centralized strategies, where a central node compiles all the information from other robots, performs the computations, and propagates the processed information or decisions to the other nodes, have several drawbacks. The whole system can fail if the central node fails, leader selection algorithms may be needed, and communication of all agents with the central system may be required. On the other hand, distributed systems are naturally more robust to individual failures since all robots play the same role. They also consider more realistic situations where agents cannot communicate with all other robots at every time instant, but instead they exchange data only with a limited number of other robots, e.g., agents within a specific distance. These situations can be best modeled using communication graphs, where nodes are the agents and edges represent communication capabilities between the robots. Additionally, since agents are moving, the topology of the graph may vary along the time, given rise to switching topologies. We analyze map merging and localization solutions for robotic systems with range limited communication, and where the computations are distributed among the robots. We also study the problem of distributed data association. We consider that a strategy is distributed when

- it does not rely on any particular communication topology and it is robust to changes in the topology;
- every robot in the team computes and obtains the global information;
- every robot plays the same role, making the system robust to individual failures;
- information is exchanged exclusively between neighbors.

1.2 Classical Approaches

Multi-robot systems have been deeply researched during the last years. A general overview of the achieved results, and the current and future research lines can be found in [28, 37]. We provide here the following references for the rendezvous [14], the deployment and coverage [22], and the formation control problems [11, 38], as some examples within the variety of different existing works. The consensus or averaging problem has a special relevance in multi-robot systems and it is connected to diverse applications, including sensor fusion, flocking, formation control, or rendezvous [27]. Several ideas presented along this document are built on consensus results in the books [4, 31].

Many existent solutions for single robot perception have been extended to multi-robot scenarios under centralized schemes, full communication between the robots, or broadcasting methods. Particle filters have been generalized to multi-robot systems assuming that the robots broadcast their controls and their observations [17]. In [32] a single global map is updated by all the robots. Robots look for coincident features in the global map that they have locally observed along the exploration. Robots update the map establishing implicit measurements for the coincident features (the difference between the Cartesian coordinates of equal features must be zero). These implicit measurements can be used as well for merging two local maps that have been previously aligned using robot-to-robot measurements [44]. Several methods [12, 30], represent maps with graphs, where nodes are local submaps [41] or laser scans [12, 30] and edges describe relative positions between the nodes. Each robot builds a new node and transmits it by broadcasts or to a central agent. After this, global optimization techniques may be applied to obtain the global metric map. The same solution could be applied to many existing submap approaches [29]. The previous methods require that each robot has the capability to communicate with all other robots at every time instant or with a central agent, i.e., they impose centralized scenarios. We are instead interested in distributed scenarios due to their robustness to robot or link failures, and due to their natural capability to operate under limited communication.

Distributed estimation methods [1, 9, 16, 18, 24, 26] maintain a joint estimate of a system that evolves with time by combining noisy observations taken by the sensor network. Early approaches sum the measurements from the different agents in IF (Information Filter) form. If the network is complete [24], then the resulting estimator is equivalent to the centralized one. In general networks the problems of cyclic updates or double counting information appear when nodes sum the same piece of data more than once. The use of the channel filter [16] avoids these problems in networks with a tree structure. The Covariance Intersection method [18] produces consistent but highly conservative estimates in general networks. More recent approaches [1, 9, 26] use distributed consensus filters to average the measurements taken by the nodes. The interest of distributed averaging is that the problems of double counting of information and cyclic updates are avoided. They, however, suffer from the delayed data problem that takes place when the nodes execute the state prediction without having incorporated all the measurements available at the current step [7]. For general communication schemes, the delayed data problem leads to an approximate KF (Kalman Filter) estimator. An interesting solution is given in [26] but its convergence is proved in the absence of observation and system noises. In the algorithm proposed in [9], authors prove that the nodes' estimates are consistent, although these estimates have disagreement. Other algorithms have been proposed that require the previous offline computation of the gains and weights of the algorithm [1]. The main limitation of all the previous works is that they consider linear systems without inputs, and where the evolution of the system is known by all the robots. The algorithms presented here can be applied to a wider class of systems, without the previous restrictions.

A related scenario are sensor fusion systems [5, 6, 21, 42, 43], where measurements acquired by several sensors are fused in a distributed fashion. Sensor fusion systems differ from the distributed perception scenario that we consider in this book in several aspects. First, sensor fusion approaches consider that the successive measurements, in our case local maps, from the same robot must be independent. In a map merging scenario this does not hold, since the local map of a robot is an evolution of any of its previous maps. Second, sensors usually observe a set of variables which are a priori known by the sensor network, e.g., temperature, humidity, etc. However, in distributed perception scenarios, robots discover elements in the environment dynamically, as they operate. Thus, it is not possible to predict which elements will be detected and inform the robot team of these elements before starting the exploration. In addition, distributed perception methods must address specific challenges such as associating the elements observed by the robots in a globally consistent way, or computing the relative poses of the robots and establishing a common reference frame for the whole robot team.

The data association problem consists of establishing correspondences between different measurements or estimates of a common element. Traditional data association methods, like the Nearest Neighbor and Maximum Likelihood [15, 19, 44], the Joint Compatibility Branch and Bound (JCBB) [25], or the Combined Constraint Data Association [3] are designed for single robot systems. They operate on two sets of elements, one containing the feature estimates and the other one containing the current observations. Multi-robot approaches have not fully addressed the problem of data association. Many methods rely on broadcasting controls and observations or submaps, see, e.g., [12, 15, 17, 30], and solve the data association using a cycle-free order, thus essentially reducing the problem to that of the single robot scenario. However, in a distributed map merging scenario like the one considered in this book, the robots may fuse their maps with any other robot's map in any order and at any time. Therefore, it is not possible to force a specific order for solving the data association and coordinated strategies are required.

The problem of localization estimation in multi-robot systems is related to the establishment of a common reference frame for the team of robots. In general, the robots start at unknown locations and do not know their relative poses. This information can be recovered by comparing their local maps and looking for overlapping regions [8, 39]. Alternatively, robots can explicitly measure their relative poses [44, 45] without the need of having overlapping regions, or even having maps. The previous methods give the relative position of a pair of robots. After that, a distributed method is required to let the robots agree on a global reference frame and obtain their positions in this frame. This problem is known as distributed network localization. Several network localization algorithms rely on range-only [2], or bearing-only [36] relative measurements of positions. Alternatively, each agent can locally combine its observations and build an estimate of the relative full-position of its neighbors using the approach described in [40] for 3D scenarios. When full-position measurements are available, the localization problem becomes linear and can thus be solved using linear optimization methods [34]. There exist works that compute not only the agents' positions but also their orientations, [13], and that track the agents' poses [20]. It is also possible to use a position estimation algorithm combined with an attitude synchronization [23, 35] or a motion coordination [10] strategy to previously align the robot orientations. Cooperative localization methods [33] take into account the noisy nature of the relative pose measurements. Here, a robotic team moves along an environment while estimating their poses. Most of the time, each robot relies on its proprioceptive measurements. When two robots meet, they obtain a noisy measurement of their relative pose and update their estimates accordingly. During this rendezvous, each robot must be able to communicate with all the other robots in the team in order to update its estimate.

1.3 Document Organization

Along this section, we have briefly introduced and revised the state of the art of the issues that appear in distributed perception scenarios for map merging and localization. In the remaining of the book, we discuss each topic in detail and propose solutions to all these issues, with a special interest in robots equipped with cameras. For each topic we provide formal proofs of the performance of the algorithms in the previous scenarios, as well as simulation results. In addition, we show validation proofs of the algorithms under real data. This document is organized as follows:

This chapter introduces the problem addressed in this document. It comments the state of the art and outlines the organization of the book.

Chapter 2 discusses the data association problem in the context of distributed robot teams with limited communication. We explain how the robots can establish correspondences between their features and the ones observed by other team members. Robots compute the associations with their neighbors using classical matching methods. We analyze distributed algorithms that allow each robot to propagate this local data and obtain the global data association relating its features with the ones of all the other robots in the network. In addition, robots identify and correct associations which are inconsistent in the global system.

Chapter 3 analyzes the localization problem for different scenarios, and presents some methods for reaching a consensus on the global frame for the robot team, using relative robot-to-robot measurements. This global frame will be used by the robot team during their operation. Robots compute the relative position of their neighbors using classical methods. From this information, they build the global frame and compute their positions in this frame in a distributed fashion. We study the use of an anchor node to define the global frame as well as methods based on the centroid.

Chapter 4 explains how to merge the information acquired by each robot in the network to build a global representation of the environment. Robots explore the environment and, simultaneously, fuse their local maps and build the global map. Therefore, robots have a representation of the environment beyond its local map during all their operation. The fusion of the local observations of all the team members leads to a merged map that contains more precise information and more features. To explain the problem and the algorithms we will consider that the ground truth data association is available. We further assume that all the robots share a common reference frame and that they know their pose in this frame.

Chapter 5 evaluates some of the presented algorithms in real scenarios. Within each chapter, we include a brief discussion of the studied algorithms with simulations.

Chapter 6 presents the conclusions of this document. The document finishes with two appendices which briefly revise the Metropolis weights and which provide some auxiliary results for distributed localization.

References

1. P. Alriksson, A. Rantzer, Distributed Kalman filtering using weighted averaging, in *International Symposium on Mathematical Theory of Networks and Systems*, Kyoto, Japan (2006)
2. B.D.O. Anderson, I. Shames, G. Mao, B. Fidan, Formal theory of noisy sensor network localization. SIAM J. Discrete Math. **24**(2), 684–698 (2010)
3. T. Bailey, E.M. Nebot, J.K. Rosenblatt, H. Durrant-Whyte, Data association for mobile robot navigation: a graph theoretic approach, in *IEEE International Conference on Robotics and Automation*, San Francisco, USA (2000), pp. 2512–2517
4. F. Bullo, J. Cortes, S. Martinez. *Distributed Control of Robotic Networks*. Applied Mathematics Series (Princeton University Press, Princeton, 2009). Electronically available at http://coordinationbook.info
5. G. Calafiore, Distributed randomized algorithms for probabilistic performance analysis. Syst. Control Lett. **58**(3), 202–212 (2009)

6. G. Calafiore, F. Abrate, Distributed linear estimation over sensor networks. Int. J. Control **82**(5), 868–882 (2009)
7. R. Carli, A. Chiuso, L. Schenato, S. Zampieri, Distributed Kalman filtering based on consensus strategies. IEEE J. Sel. Areas Commun. **26**, 622–633 (2008)
8. S. Carpin, Fast and accurate map merging for multi-robot systems. Auton. Robots **25**(3), 305–316 (2008)
9. D.W. Casbeer. R.Beard. Distributed information filtering using consensus filters, in *American Control Conference*, St. Louis, USA (2009), pp. 1882–1887
10. J. Cortes, Global and robust formation-shape stabilization of relative sensing networks. Automatica **45**(12), 2754–2762 (2009)
11. W.B. Dunbar, R.M. Murray, Distributed receding horizon control for multi-vehicle formation stabilization. Automatica **42**(4), 549–558 (2006)
12. D. Fox, J. Ko, K. Konolige, B. Limketkai, D. Schulz, B. Stewart, Distributed multirobot exploration and mapping. IEEE Proc. **94**(7), 1325–1339 (2006)
13. M. Franceschelli, A. Gasparri, On agreement problems with gossip algorithms in absence of common reference frames, in *IEEE International Conference on Robotics and Automation* (Anchorage, USA, 2010), pp. 4481–4486
14. A. Ganguli, J. Cortés, F. Bullo, Multirobot rendezvous with visibility sensors in nonconvex environments. IEEE Trans. Robot. **25**(2), 340–352 (2009)
15. A. Gil, O. Reinoso, M. Ballesta, M. Julia, Multi-robot visual SLAM using a rao-blackwellized particle filter. Robot. Auton. Syst. **58**(1), 68–80 (2009)
16. S. Grime, H.F. Durrant-Whyte, Data fusion in decentralized sensor networks. Control Eng. Pract. **2**(5), 849–863 (1994)
17. A. Howard, Multi-robot simultaneous localization and mapping using particle filters. Int. J. Robot. Res. **25**(12), 1243–1256 (2006)
18. S. Julier, J.K. Uhlmann, General decentralised data fusion with covariance intersection (CI), in *Handbook of Multisensor Data Fusion*, ed. by D.L. Hall, J. Llinas (CRC Press, Boca Raton, 2001)
19. M. Kaess, F. Dellaert, Covariance recovery from a square root information matrix for data association. Robot. Auton. Syst. **57**(12), 1198–1210 (2009)
20. J. Knuth, P. Barooah, Distributed collaborative localization of multiple vehicles from relative pose measurements, in *Allerton Conference on Communications, Control and Computing* (Urbana-Champaign, USA, October 2009), pp. 314–321
21. K.M. Lynch, I.B. Schwartz, P. Yang, R.A. Freeman, Decentralized environmental modeling by mobile sensor networks. IEEE Trans. Robot. **24**(3), 710–724 (2008)
22. S. Martínez, F. Bullo, J. Cortés, E. Frazzoli, On synchronous robotic networks—part II: Time complexity of rendezvous and deployment algorithms, in *IEEE Conference on Decision and Control* (Seville, Spain, 2005), pp. 8313–8318
23. N. Mostagh, A. Jadbabaie, Distributed geodesic control laws for flocking of nonholonomic agents. IEEE Trans. Autom. Control **52**(4), 681–686 (2007)
24. E.M. Nebot, M. Bozorg, H.F. Durrant-Whyte, Decentralized architecture for asynchronous sensors. Auton. Robots **6**(2), 147–164 (1999)
25. J. Neira, J.D. Tardós, Data association in stochastic mapping using the joint compatibility test. IEEE Trans. Robot. Autom. **17**(6), 890–897 (2001)
26. R. Olfati-Saber, Distributed Kalman filtering for sensor networks, in *IEEE Conference on Decision and Control* (2007), pp. 5492–5498
27. R. Olfati-Saber, R.M. Murray, Consensus problems in networks of agents with switching topology and time-delays. IEEE Trans. Autom. Control **49**(9), 1520–1533 (2004)
28. L. Parker, Distributed intelligence: overview of the field and its application in multi-robot systems. J. Phys. Agents **2**(1), 5–14 (2008)
29. L.M. Paz, J.D. Tardos, J. Neira, Divide and conquer: EKF SLAM in $O(n)$. IEEE Trans. Robot. **24**(5), 1107–1120 (2008)
30. M. Pfingsthorn, B. Slamet, A. Visser, in *A Scalable Hybrid Multi-robot SLAM Method for Highly Detailed Maps*, vol. 5001, Lecture Notes in Artificial Intelligence, ed. by U. Visser, F. Ribeiro, T. Ohashi, F. Dellaert (2008), pp. 457–464

31. W. Ren, R.W. Beard, *Distributed Consensus in Multi-vehicle Cooperative Control, Communications and Control Engineering* (Springer, London, 2008)
32. D. Rodríguez-Losada, F. Matía, A. Jiménez, Local maps fusion for real time multirobot indoor simultaneous localization and mapping, in *IEEE International Conference on Robotics and Automation*, New Orleans, USA (2004) pp. 1308–1313
33. S.I. Roumeliotis, G.A. Bekey, Distributed multirobot localization. IEEE Trans. Robot. Autom. **18**(5), 781–795 (2002)
34. W.J. Russell, D. Klein, J.P. Hespanha, Optimal estimation on the graph cycle space, in *American Control Conference*, Baltimore, USA (2010), pp. 1918–1924
35. A. Sarlette, R. Sepulchre, N.E. Leonard, Autonomous rigid body attitude synchronization. Automatica **45**(2), 572–577 (2008)
36. A. Savvides, W.L. Garber, R.L. Moses, M.B. Srivastava, An analysis of error inducing parameters in multihop sensor node localization. IEEE Trans. Mobile Comput. **4**(6), 567–577 (2005)
37. A.C. Schultz, L.E. Parker (eds.), *Multi-Robot Systems: From Swarms to Intelligent Automata* (Kluwer Academic Publishers, Dordrecht, 2002)
38. H.G. Tanner, G.J. Pappas, V. Kumar, Leader-to-formation stability. IEEE Trans. Robot. Autom. **20**(3), 443–455 (2004)
39. S. Thrun, Y. Liu, Multi-robot SLAM with sparse extended information filters, in *International Symposium of Robotics Research*, Sienna, Italy (2003), pp. 254–266
40. N. Trawny, X.S. Zhou, K.X. Zhou, S.I. Roumeliotis, Inter-robot transformations in 3-d. IEEE Trans. Robot. **26**(2), 226–243 (2010)
41. S.B. Williams, H. Durrant-Whyte, Towards multi-vehicle simultaneous localisation and mapping, in *IEEE International Conference on Robotics and Automation*, Washington (2002), pp. 2743–2748
42. L. Xiao, S. Boyd, Fast linear iterations for distributed averaging. Syst. Control Lett. **53**, 65–78 (2004)
43. L. Xiao, S. Boyd, S. Lall, A space-time diffusion scheme for peer-to-peer least-square estimation, in *Symposium on Information Processing of Sensor Networks (IPSN)*, Nashville, TN (2006), pp. 168–176
44. X.S. Zhou, S.I. Roumeliotis. Multi-robot SLAM with unknown initial correspondence: The robot rendezvous case, in *IEEE/RSJ International Conference on Intelligent Robots and Systems*, Beijing, China (2006), pp. 1785–1792
45. X.S. Zhou, S.I. Roumeliotis, Robot-to-robot relative pose estimation from range measurements. IEEE Trans. Robot. **24**(6), 1379–1393 (2008)

Chapter 2
Distributed Data Association

Abstract In this chapter, we address the association of features observed by the robots in a network with limited communications. At every time instant, each robot can only exchange data with a subset of the robot team that we call its neighbors. Initially, each robot solves a local data association with each of its neighbors. After that, the robots execute the proposed algorithm to agree on a data association between all their local observations. One inconsistency appears when chains of local associations give rise to two features from one robot being associated among them. In finite time, the algorithm finishes with a data association which is free of inconsistent matches. We show the performance of the proposed algorithms through simulations. Experiments with real data can be found in the last chapter.

Keywords Data association · Limited communication · Distributed systems · Parallel computation

2.1 Introduction

In multi-robot systems, a team of robots cooperatively perform some task in a more efficient way than a single robot would do. In this chapter, we address the data association problem. It consists of establishing correspondences between different measurements or estimates of a common element. It is of high interest in localization, mapping, exploration, and tracking applications [4]. There exists a wide variety of matching functions. The Nearest Neighbor (NN), and the Maximum Likelihood (ML), are widely used methods which associate each observation with its closest feature in terms of the Euclidean or the Mahalanobis distance [13, 15, 24]. Other popular method is the Joint Compatibility Branch and Bound (JCBB) [19], which considers the compatibility of many associations simultaneously. The combined constraint data association [5] builds a graph where the nodes are individually compatible associations and the edges relate binary compatible assignments. Over this graph, a maximal common subgraph problem is solved for finding the maximum clique in the graph. Scan matching and iterative closest point (ICP) [8] are popular methods for comparing two laser scans. Other methods, like the multiple hypothesis tracking, and the joint probabilistic data association, maintain many association hypothesis instead

© The Author(s) 2015
R. Aragues et al., *Parallel and Distributed Map Merging and Localization*,
SpringerBriefs in Computer Science, DOI 10.1007/978-3-319-25886-7_2

of selecting one of them. And there exists many variations of these techniques that combine RANSAC [11] for higher robustness. All these matching functions operate on elements from two sets. One set usually contains the current observations, and the other one consists of the feature estimates. These sets may be two images, two laser scans, or two probabilistic maps.

Lately, many localization, mapping, and exploration algorithms for multi-robot systems have been presented. However, they have not fully addressed the problem of multi-robot data association. Some solutions have been presented for merging two maps [22, 24] that do not consider a higher number of robots. Many approaches rely on broadcasting all controls and observations measured by the robots. Then, the data association is solved like in a single robot scenario, using scan matching and ICP for laser scans [12, 14, 16, 21], or NN, ML, and visual methods for feature-based maps [13, 17]. Solutions based on submaps usually transform one of them into an observation of another. The local submaps are merged with the global map following a sequence [23], or in a hierarchical binary tree fashion [7].

In these methods, the problem of inconsistent data associations is avoided by forcing a cycle-free merging order. This limitation has also been detected in the computer vision literature. In [10] they approach an inconsistent association problem for identifying equal regions in different views. They consider a centralized scenario, where each two views are compared among them in a 2-by-2 way. Then, their results are arranged on a graph where associations are propagated and conflicts are solved. The work in [9], from the target tracking literature, simultaneously considers the association of all local maps. It uses an expectation-maximization method for both, computing the data association and the final global map. The main limitation of this work is that the data from all sensors needs to be processed together, what implies a centralized scheme, or a broadcast method.

All the previous methods rely on centralized schemes, full communication between the robots, or broadcasting methods. However, in multi-robot systems, distributed approaches are more interesting. They present a natural robustness to individual failures since there are no central nodes. Besides, they do not rely on any particular communication scheme, and they are robust to changes in the topology. On the other hand, distributed algorithms introduce an additional level of complexity in the algorithm design. Although, the robots make decisions based on their local data, the system must exhibit a global behavior.

In this chapter, we address the data association problem for distributed robot systems. Each of our robots posse a local observation of the environment. Instead of forcing a specific order for associating their observations, we allow the robots to compute its data association with each of its neighbors in the graph. Although this scenario is more flexible, it may lead to inconsistent global data associations in the presence of cycles in the communication graph. These inconsistencies are detected when chains of local associations give rise to two features from one robot being associated among them. These situations must be correctly identified and solved before merging the data. Otherwise, the merging process would be wrong and could not be undone. We approach this problem under limited communications. So, instead of comparing any two local observations among them, only the local observations of neighboring robots can be compared. Besides, there is no central node that has knowledge of all the local associations and each robot exclusively knows the associations computed by itself. Then, each robot updates its local information by communicating with its neighbors. We present an algorithm where, finally, each robot is capable of detecting and solving any inconsistent association that involves any of its features.

2.2 Problem Description

We consider, a robotic team composed of $n \in \mathbb{N}$ robots. The n robots have communication capabilities to exchange information with the other robots. However, these communications are limited. Let $\mathcal{G}_{com} = (\mathcal{V}_{com}, \mathcal{E}_{com})$ be the undirected communication graph. The nodes are the robots, $\mathcal{V}_{com} = \{1, \ldots, n\}$. If two robots i, j can exchange information then there is an edge between them, $(i, j) \in \mathcal{E}_{com}$. Let $\mathcal{N}_i$ be the set of neighbors of robot i,

$$\mathcal{N}_i = \{j \mid (i, j) \in \mathcal{E}_{com}\}.$$

Each robot i has observed a set $\mathcal{S}_i$ of m_i features,

$$\mathcal{S}_i = \{f_1^i, \ldots, f_{m_i}^i\}.$$

It can compute the local data association between its own set $\mathcal{S}_i$, and the sets of its neighbors $\mathcal{S}_j$, with $j \in \mathcal{N}_i$. However, these data associations are not perfect. There may appear inconsistent data associations relating different features from the same set $\mathcal{S}_i$ (Fig. 2.1). If the robots merge their data as soon as they solve the local data association, inconsistent associations cannot be managed since the merging cannot be undone. The goal of the algorithm is to detect and resolve these inconsistent associations before executing the merging.

In order to make the reading easy, along the chapter we use the indices i, j, and k to refer to robots and indices r, r', s, s', to refer to features. The rth feature observed by the ith robot is denoted as f_r^i. Given, a matrix A, the notations $A_{r,s}$ and $[A]_{r,s}$

Fig. 2.1 Robots $A, B, C,$ and D associate their features comparing their maps in a two-by-two way. As a result, there is a path (*dashed line*) between f_1^D and f_2^D. This is an inconsistent association

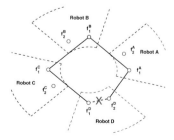

correspond to the (r, s) entry of the matrix, whereas, A_{ij} denotes the (i, j) block when the matrix is defined by blocks. We let $\mathbf{I}_k$ be the $k \times k$ identity matrix, and $\mathbf{0}_{k_1 \times k_2}$ a $k_1 \times k_2$ matrix with all entries equal to zero.

2.2.1 Matching Between Two Cameras

Let F be a function that computes the local data association between any two sets of features, $\mathscr{S}_i$ and $\mathscr{S}_j$, and returns an association matrix $F(\mathscr{S}_i, \mathscr{S}_j) = \mathbf{A}_{ij}$ where $\mathbf{A}_{ij} \in \mathbb{N}^{m_i \times m_j}$,

$$[\mathbf{A}_{ij}]_{r,s} = \begin{cases} 1 & \text{if } f_r^i \text{ and } f_s^j \text{ are associated,} \\ 0 & \text{otherwise,} \end{cases}$$

for $r = 1, \ldots, m_i$ and $s = 1, \ldots, m_j$. We assume that F satisfies the following conditions.

Assumption 1 (*Self Association*) When F is applied to the same set $\mathscr{S}_i$, it returns the identity, $F(\mathscr{S}_i, \mathscr{S}_i) = \mathbf{A}_{ii} = \mathbf{I}$. □

Assumption 2 (*Unique Association*) The returned association $\mathbf{A}_{ij}$ has the property that the features are associated in a one-to-one way,

$$\sum_{r=1}^{m_i} [\mathbf{A}_{ij}]_{r,s} \leq 1 \text{ and } \sum_{s=1}^{m_j} [\mathbf{A}_{ij}]_{r,s} \leq 1,$$

for all $r = 1, \ldots, m_i$ and $s = 1, \ldots, m_j$. □

Assumption 3 (*Symmetric Association*) Robots i and j associate their features in the same way. Given two sets $\mathscr{S}_i$ and $\mathscr{S}_j$ it holds that $F(\mathscr{S}_i, \mathscr{S}_j) = \mathbf{A}_{ij} = \mathbf{A}_{ji}^T = (F(\mathscr{S}_j, \mathscr{S}_i))^T$. □

Additionally, the local matching function may give information of the quality of each associations. The management of this information is discussed in Sect. 2.6.

We do not make any assumptions about the sets of features used by the cameras. However, we point out that the better the initial matching is, the better the global matching will be.

2.2.2 Centralized Matching Between n Cameras

Let us consider now the situation in which there are n cameras and a central unit with the n sets of features available. In this case, F can be applied to all the pairs of sets of features, $\mathscr{S}_i, \mathscr{S}_j$, for $i, j \in \{1, \ldots, n\}$. The results of all the associations can

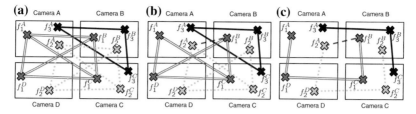

Fig. 2.2 Different association graphs. **a** Centralized matching with perfect association function. The graph is formed by disjoint cliques. **b** Centralized matching with imperfect association. Some links are missed, (f_1^A, f_1^B) and (f_2^A, f_2^B), and spurious links appear, (f_2^A, f_1^B). As a consequence, a subset of the features form a *conflictive set*. **c** Matching with limited communications. Now, the links between A and C, and B and D cannot be computed because they are not neighbors in $\mathscr{G}_{com}$. Moreover, the information available to each camera is just the one provided by its neighbors

be represented by an undirected graph $\mathscr{G}_{cen} = (\mathscr{F}_{cen}, \mathscr{E}_{cen})$. Each node in $\mathscr{F}_{cen}$ is a feature f_r^i, for $i = 1, \ldots, n, r = 1, \ldots, m_i$. There is an edge between two features f_r^i, f_s^j iff $[\mathbf{A}_{ij}]_{r,s} = 1$.

For a perfect matching function, the graph $\mathscr{G}_{cen}$ exclusively contains disjoint cliques, identifying features observed by multiple cameras (Fig. 2.2a). However, in real situations, the matching function will miss some matches and will consider as good correspondences some spurious matches (Fig. 2.2b). As a consequence, inconsistent associations relating different features from the same set $\mathscr{S}_i$ may appear.

Definition 1 An *association set* is a set of features such that they form a connected component in $\mathscr{G}_{cen}$. Such set is a *conflictive set* or an *inconsistent association* if there exists a path in $\mathscr{G}_{cen}$ between two or more features observed by the same camera. A feature is *inconsistent* or *conflictive* if it belongs to an inconsistent association. □

Centralized solutions to overcome this problem are found in [3]. The latter one is also well suited for a distributed implementation but yet requires that any pair of images can be matched. In camera networks this implies global communications, which is not always possible.

2.2.3 Distributed Matching Between n Cameras

Let us consider now that there is no central unit with all the information and there are n robots, each one with a camera and a process unit with limited communication capabilities. The robots are scattered forming a network with communications described with the undirected communication graph $\mathscr{G}_{com} = (\mathscr{V}_{com}, \mathscr{E}_{com})$ introduced at the beginning of this section.

In this case, due to communication restrictions, local matches can only be found within direct neighbors. As a consequence, the matching graph computed in this situation will be a subgraph of the centralized one, $\mathscr{G}_{dis} = (\mathscr{F}_{dis}, \mathscr{E}_{dis}) \subseteq \mathscr{G}_{cen}$,

(Fig. 2.2c). It has the same set of nodes, $\mathscr{F}_{dis} = \mathscr{F}_{cen}$, but it has an edge between two features f_r^i, f_s^j only if the edge exists in $\mathscr{G}_{cen}$ and the robots i and j are neighbors in the communication graph,

$$\mathscr{E}_{dis} = \{(f_r^i, f_s^j) \mid (f_r^i, f_s^j) \in \mathscr{E}_{cen} \wedge (i, j) \in \mathscr{E}_{com}\}.$$

Along this chapter, we name m_{sum} the number of features, $|\mathscr{F}_{dis}| = \sum_{i=1}^{n} m_i = m_{sum}$. We name d_f the diameter of $\mathscr{G}_{dis}$, the length of the longest path between any two nodes in $\mathscr{G}_{dis}$, and we name d_v the diameter of the communication graph, $\mathscr{G}_{com}$. The diameters satisfy $d_f \leq m_{sum}$ and $d_v \leq n$. We name $\mathbf{A} \in \mathbb{N}^{m_{sum} \times m_{sum}}$ the adjacency matrix of $\mathscr{G}_{dis}$,

$$\mathbf{A} = \begin{bmatrix} \mathbf{A}_{11} & \dots & \mathbf{A}_{1n} \\ \vdots & \ddots & \vdots \\ \mathbf{A}_{n1} & \dots & \mathbf{A}_{nn} \end{bmatrix}, \tag{2.1}$$

where

$$\mathbf{A}_{ij} = \begin{cases} F(\mathscr{S}_i, \mathscr{S}_j) & \text{if } j \in \{\mathscr{N}_i \cup i\}, \\ \mathbf{0} & \text{otherwise.} \end{cases} \tag{2.2}$$

Let us note that in this case none of the robots has the information of the whole matrix. Robot i has only available the submatrix corresponding to its own local matches $\mathbf{A}_{ij}$, $j = 1, \dots, n$. Under these circumstances the problem is formulated as follows: Given a network with communications defined by a graph, $\mathscr{G}_{com}$, and an association matrix $\mathbf{A}$ scattered over the network, find the global matches and the possible inconsistencies in a distributed way. In case there are conflicts, find alternative associations free of them.

2.3 Propagation of Local Associations

Considering Definition 1 (Sect. 2.2.2), we observe that in order to find the data association sets with the relationship between the features observed by the different robots, it is required to compute the paths that exist among the elements in $\mathscr{G}_{dis}$. We show a process where robots start considering their local matches, and incrementally they propagate these local matches and discover all the paths between the features observed by the robot team. This information allows them as well to detect inconsistent associations (Definition 1). As the following lemma states [6], given a graph $\mathscr{G}_{dis}$, the powers of its adjacency matrix contains the information about the number of paths existing between the nodes of $\mathscr{G}_{dis}$:

Lemma 1 (Lemma 1.32 [6]) *Let $\mathscr{G}_{dis}$ be a weighted graph of order $|\mathscr{V}|$ with unweighted adjacency matrix $A \in \{0, 1\}^{|\mathscr{V}| \times |\mathscr{V}|}$, and possibly with self loops. For all*

$i, j \in \{1, \ldots, |\mathcal{V}|\}$ and $t \in \mathbb{N}$ the (i, j) entry of the tth power of $\mathbf{A}$, $\mathbf{A}^t$, equals the number of paths of length t (including paths with self-loops) from node i to node j.

Algorithm 1 The computation of the powers of $\mathbf{A}$ requires, a priori, the information about the whole matrix. We show now that this computation can also be done in a distributed manner [1]. Let each robot $i \in \mathcal{V}_{com}$ maintain the blocks within $\mathbf{A}^t$ associated to its own features, $X_{ij}(t) \in \mathbb{N}^{m_i \times m_j}$, $j = 1, \ldots, n$, $t \geq 0$, which are initialized as

$$X_{ij}(0) = \begin{cases} \mathbf{I}, & j = i, \\ \mathbf{0}, & j \neq i, \end{cases} \tag{2.3}$$

and are updated, at each time step, with the following algorithm

$$X_{ij}(t+1) = \sum_{k \in \{\mathcal{N}_i \cup i\}} \mathbf{A}_{ik} X_{kj}(t), \tag{2.4}$$

with $\mathbf{A}_{ik}$ as defined in (2.2). It is observed that the algorithm is fully distributed because the robots only use information about its direct neighbors in the communication graph.

Theorem 1 Let $[\mathbf{A}^t]_{ij} \in \mathbb{N}^{m_i \times m_j}$ be the block within $\mathbf{A}^t$ related to the associations between robot i and robot j. The matrices $X_{ij}(t)$ computed by each robot i using the distributed algorithm (2.4) are exactly the submatrices $[\mathbf{A}^t]_{ij}$,

$$X_{ij}(t) = [\mathbf{A}^t]_{ij}, \tag{2.5}$$

for all $i, j \in \{1, \ldots, n\}$ and all $t \in \mathbb{N}$.

Proof The proof is done using induction. First, we show that Eq. (2.5) is satisfied for $t = 0$. In this case, we have that $\mathbf{A}^0 = \mathbf{I}$, thus for all $i, j \in \{1, \ldots, n\}$, $[\mathbf{A}^0]_{ii} = \mathbf{I}$ and $[\mathbf{A}^0]_{ij} = \mathbf{0}$, which is exactly the initial value of the variables X_{ij} (Eq. (2.3)).
Now we have that for any $t > 0$,

$$[\mathbf{A}^t]_{ij} = \sum_{k=1}^{n} \mathbf{A}_{ik}[\mathbf{A}^{t-1}]_{kj} = \sum_{k \in \{\mathcal{N}_i \cup i\}} \mathbf{A}_{ik}[\mathbf{A}^{t-1}]_{kj},$$

because $\mathbf{A}_{ik} = \mathbf{0}$ for $k \notin \{\mathcal{N}_i \cup i\}$. Assuming that for all $i, j \in \{1, \ldots, n\}$ and a given $t > 0$, $X_{ij}(t-1) = [\mathbf{A}^{t-1}]_{ij}$ is true, then

$$X_{ij}(t) = \sum_{k \in \{\mathcal{N}_i \cup i\}} \mathbf{A}_{ik} X_{kj}(t-1) = \sum_{k \in \{\mathcal{N}_i \cup i\}} \mathbf{A}_{ik}[\mathbf{A}^{t-1}]_{kj} = [\mathbf{A}^t]_{ij}.$$

Then, by induction, $X_{ij}(t) = [\mathbf{A}^t]_{ij}$ is true for all $t > 0$. $\qquad\square$

Corollary 1 *The variables $X_{ij}(t)$ contain the information about all the paths of length t between features observed by robots i and j.*

Proof By direct application of Lemma 1.   □

Analyzing the previous algorithm the first issue to deal with is how to simplify the computation of the matrices in order to avoid high powers of $\mathbf{A}$. In the case, we are studying it is just required to know if there is a path between two elements in $\mathscr{G}_{dis}$ and not how many paths are. This means that in this situation it is enough that $[X_{ij}(t)]_{r,s} > 0$ in order to know that features f_r^i and f_s^j are connected by a path. Another issue is to decide when the algorithm in (2.4) must stop. Since the maximum length of a path between any two nodes in a graph is its diameter, then after d_f iterations the algorithm should stop. However, in general situations the robots will not know neither d_f nor m_{sum}, which makes this decision hard to be made a priori.

Definition 2 We will say that two matrices $\mathbf{A}$ and $\bar{\mathbf{A}}$ of the same dimensions are equivalent, $\mathbf{A} \sim \bar{\mathbf{A}}$, if for all r and s it holds

$$[\mathbf{A}]_{r,s} > 0 \Leftrightarrow [\bar{\mathbf{A}}]_{r,s} > 0 \text{ and } [\mathbf{A}]_{r,s} = 0 \Leftrightarrow [\bar{\mathbf{A}}]_{r,s} = 0. \qquad \square$$

In practice any equivalent matrix to the $X_{ij}(t)$ will provide the required information, which allows to simplify the computations simply by changing any positive value in the matrices by 1. Moreover, the equivalency is also used to find a criterion to stop the algorithm:

Proposition 1 *For a robot i, let t_i be the first time instant, t, such that $X_{ij}(t) \sim X_{ij}(t-1)$ for all $j = 1, \ldots, n$. Then robot i can stop to execute the algorithm at time t_i.*

Proof Let $\bar{X}_{ij}(t)$ be the components in $X_{ij}(t)$, such that $[X_{ij}(t-1)]_{r,s} = 0$ and $[X_{ij}(t)]_{r,s} > 0$. The cardinal, $|\bar{X}_{ij}(t)|$, represents the number of features $f_s^j \in \mathscr{S}_j$ such that the minimum path length in $\mathscr{G}_{dis}$ between them and one feature $f_r^i \in \mathscr{S}_i$ is t. At time t_i, $X_{ij}(t_i) \sim X_{ij}(t_i - 1)$ $\forall j$ for the first time, and then $\sum_{j=1}^n |\bar{X}_{ij}(t_i)| = 0$ because no component has changed its value from zero to a positive. This means that there is no path of minimum distance t_i linking any feature f_r^i with any other feature in $\mathscr{G}_{dis}$. By the physical properties of a path, it is obvious that if there are no features at minimum distance t_i, it will be impossible that a feature is at minimum distance $t_i + 1$ and all the paths that connect features of robot i with any other feature have been found.   □

Corollary 2 *All the robots end the execution of the iteration rule (2.4) in at most in $d_f + 1$ iterations.*

Proof Recalling that the maximum distance between two nodes in $\mathscr{G}_{dis}$ is the diameter of the graph, denoted by d_f, then $\sum_{j=1}^n |\bar{X}_{ij}(d_f + 1)| = 0$ for all $i = 1, \ldots, n$.   □

If a robot j at time t does not receive the information $X_{ij}(t)$ from robot i then it will use the last matrix received, because robot i has already finished computing its paths and $X_{ij}(t) \sim X_{ij}(t-1)$.

When the algorithm finishes, each robot i has the information about all the association paths of its features and the features of the rest of the robots in the network in the different variables $X_{ij}(t_i)$. It remains to analyze which features are conflictive and which are not.

Algorithm 2 The robots detect all the conflictive features using two simple rules. A feature f_r^i is conflictive if and only if one of the following conditions are satisfied:

(i) There exists other feature $f_{r'}^i$, with $r \neq r'$, such that

$$[X_{ii}(t_i)]_{r,r'} > 0; \tag{2.6}$$

(ii) There exist features f_s^j and $f_{s'}^j$, $s \neq s'$, such that

$$[X_{ij}(t_i)]_{r,s} > 0 \text{ and } [X_{ij}(t_i)]_{r,s'} > 0. \tag{2.7}$$

In conclusion, the proposed algorithm will be able to find all the inconsistencies in a finite number of iterations. The algorithm is distributed and it is based only on local interactions between the robots. Each robot only needs to know its local data associations. It updates its information based on the data exchanged with its neighbors. When the algorithm finishes, each robot i can extract from its own matrices $X_{ij}(t_i)$ all the information of any conflict that involves any of its features. If the robot has any conflictive feature, it also knows the rest of features that belong to the conflictive set independently of the robot that observed such features. An algorithm to carry out the same process, but exploiting local information through the use of logical operations can also be used [18].

2.4 Algorithm Based on Trees

The resolution of inconsistent associations consists of deleting edges from $\mathcal{G}_{dis}$ so that the resulting graph is conflict-free.

Definition 3 Let C denote the number of conflictive sets in $\mathcal{G}_{dis}$. We say a conflictive set $\mathcal{C}$ is *detectable* by a robot i if there exists a $r \in \{1, \dots, m_i\}$ such that $f_r^i \in C$. The set of robots that detect a conflictive set $\mathcal{C}$ is $R \subseteq \mathcal{V}_{com}$. The number of features from each robot $i \in R$ involved in $\mathcal{C}$ is $\tilde{m}_i$. We say $\mathcal{G}_{dis}$ is *conflict-free* if $C = 0$. $\square$

All the edges whose deletion transforms $\mathcal{G}_{dis}$ into a conflict-free graph, belong to any of the C conflictive sets of $\mathcal{G}_{dis}$. Since the conflictive sets are disjoint, they can be considered separately. From now on, we focus on the resolution of one of the conflictive sets $\mathcal{C}$. The other conflictive sets are managed in the same way.

The resolution problem consists of partitioning $\mathscr{C}$ into a set of disjoint conflict-free components $\mathscr{C}_q$ such that

$$\bigcup_q \mathscr{C}_q = \mathscr{C}, \text{ and } \mathscr{C}_q \cap \mathscr{C}_{q'} = \emptyset,$$

for all $q, q' = 1, 2, \ldots$. The number of such conflict-free components is a priori unknown and it will be discussed later in this section.

Obtaining an optimal partition that minimizes the number of deleted edges is complicated. If there were only two inconsistent features f_r^i, $f_{r'}^i$, it could be approached as a max-flow min-cut problem [20]. However, in general there will be more inconsistent features, $\tilde{m}_i \geq 2$, within $\mathscr{C}$ associated to a robot $i \in R$. Besides, there may also be $\tilde{m}_j \geq 2$ inconsistent features belonging to a different robot $j \in R$. The application of [20] separately to any pair of inconsistent features does not necessarily produce an optimal partition. It may happen that a single edge deletion simultaneously resolves more than one inconsistent association. Therefore, an optimal solution should consider multiple combinations of edge deletions, what makes the problem computationally intractable, and imposes a centralized scheme. The algorithm presented is not optimal but is efficient and is proven to be correct and can be applied in distributed systems.

Proposition 2 *Let R be the set of robots that detect $\mathscr{C}$. Let $i_\star$ be the root robot with the most features involved in $\mathscr{C}$,*

$$i_\star = \arg\max_{i \in R} \tilde{m}_i. \tag{2.8}$$

The number of conflict-free components in which $\mathscr{C}$ can be decomposed is lower bounded by $\tilde{m}_{i_\star}$.

Proof Each conflict-free component can contain, at most, one feature from a robot $i \in R$. Then, there must be at least, $\max_{i \in R} \tilde{m}_i = \tilde{m}_{i_\star}$ components.   ⊓

The resolution algorithm [1] constructs $\tilde{m}_{i_\star}$ conflict-free components using a strategy close to a BFS tree construction. Initially, each robot i detects the conflictive sets for which it is the root using its local information $X_{i1}(t_i), \ldots, X_{in}(t_i)$. The root robot for a conflictive set is the one with the most inconsistent features involved. In case two robots have the same number of inconsistent features, the one with the lowest robot id is selected. Then, each robot executes the resolution algorithm (Algorithm 2.4.1).

The root robot creates $\tilde{m}_{i_\star}$ components and initializes each component $\mathscr{C}_q$ with one of its features $f^{i_\star} \in \mathscr{C}$. Then, it tries to add to each component $\mathscr{C}_q$ the features directly associated to $f^{i_\star} \in \mathscr{C}_q$. Let us consider that f_s^j has been assigned to $\mathscr{C}_q$. For all f_r^i such that $[\mathbf{A}_{ij}]_{r,s} = 1$, robot j sends a component request message to robot i. When robot i receives it, it may happen that

(a) f_r^i is already assigned to $\mathscr{C}_q$;
(b) f_r^i is assigned to a different component;

Algorithm 2.4.1 Spanning Trees - Robot i

1: – *Initialization*
2: **for** each conflictive set $\mathscr{C}$ for which i is root ($i = i_\star$) **do**
3:     create $\tilde{m}_{i_\star}$ components
4:     assign each inconsistent feature $f_r^{i\star} \in \mathscr{C}$ to a different component $\mathscr{C}_q$
5:     send component request to all its neighboring features
6: **end for**
7:
8: – *Algorithm*
9: **for** each component request from f_s^j to f_r^i **do**
10:     **if** (b) or (c) **then**
11:         $[\mathbf{A}_{ij}]_{r,s} = 0$
12:         send reject message to j
13:     **else if** (d) **then**
14:         assign f_r^i to the component
15:         send component request to all its neighboring features
16:     **end if**
17: **end for**
18: **for** each component reject from f_s^j to f_r^i **do**
19:     $[\mathbf{A}_{ij}]_{r,s} = 0$
20: **end for**

(c) other feature $f_{r'}^i$ is already assigned to $\mathscr{C}_q$;

(d) f_r^i is unassigned and no feature in i is assigned to $\mathscr{C}_q$.

In case (a), f_r^i already belongs to the component $\mathscr{C}_q$ and robot i does nothing. In cases (b) and (c), f_r^i cannot be added to $\mathscr{C}_q$; robot i deletes the edge $[\mathbf{A}_{ij}]_{r,s}$ and replies with a reject message to robot j; when j receives the reject message, it deletes the equivalent edge $[\mathbf{A}_{ji}]_{s,r}$. In case (d), robot i assigns its feature f_r^i to the component $\mathscr{C}_q$ and the process is repeated.

Theorem 2 *Let us consider that each robot $i \in \mathscr{V}_{com}$ executes the distributed resolution algorithm (Algorithm 2.4.1) on $\mathscr{G}_{dis}$, obtaining $\mathscr{G}'_{dis}$,*

(i) *after $t = n$ iterations no new features are added to any component $\mathscr{C}_q$ and the algorithm finishes;*

(ii) *each obtained $\mathscr{C}_q$ is a connected component in $\mathscr{G}'_{dis}$;*

(iii) *$\mathscr{C}_q$ is conflict free;*

(iv) *$\mathscr{C}_q$ contains at least two features;*

for all $q \in \{1, \ldots, \tilde{m}_{i_\star}\}$ and all conflictive sets.

Proof (*i*) The maximal depth of a conflict-free component is n since, if there were more features, at least two of them would belong to the same robot. Then, after at most n iterations of this algorithm, no more features are added to any component $\mathscr{C}_q$ and the algorithm finishes.

(ii) There is a path in $\mathcal{G}_{dis}$ between any two features belonging to a conflictive set $\mathcal{C}$. Therefore, there is also a path in $\mathcal{G}_{dis}$ between any two features assigned to the same component $\mathcal{C}_q$. Since the algorithm does not delete edges from $\mathcal{G}_{dis}$ within a component (case (a)), then $\mathcal{C}_q$ it is also connected in $\mathcal{G}'_{dis}$. Since none feature can be assigned to more than one component (case (b)), the components are disjoint. Therefore, $\mathcal{C}_q$ is a connected component in $\mathcal{G}'_{dis}$.

(iii) By construction, two features from the same robot are never assigned to the same component $\mathcal{C}_q$ (case (c)). Therefore, each component is conflict-free.

(iv) Each conflictive set has more than one feature. Because of Assumptions 1 and 2, each feature and its neighbors are conflict free. Therefore, each component $\mathcal{C}_q$ contains, at least, its originating feature, and a neighboring feature. Thus, it has at least two features.   □

Corollary 3 *After executing Algorithm 2.4.1, the size of each conflict set $\mathcal{C}$ is reduced by at least $2\,\tilde{m}_{i_\star}$, where $\tilde{m}_{i_\star} \geq 2$.*   □

When the algorithm finishes, each original conflictive set $\mathcal{C}$ has been partitioned into $\tilde{m}_{i_\star}$ conflict-free components. It may happen that a subset of features remains unassigned. These features may still be conflictive in $\mathcal{G}'_{dis}$. The detection algorithm (Algorithm 2) can be executed on the subgraph defined by this smaller subset of features.

Proposition 3 *Consider each robot i iteratively executes the detection (Sect. 2.3) and the resolution (Sect. 2.4) algorithms. Then, in a finite number of iterations, all conflictive sets disappear.*

Proof After each execution of the resolution algorithm, the size of each conflict set $\mathcal{C}$ is reduced by, at least, $2\,\tilde{m}_{i_\star} \geq 4$ (Corollary 3). Then, in a finite number of iterations, it happens that $|\mathcal{C}| < 4$. A set with 3 features f_r^i, $f_{r'}^i$, f_s^j cannot be conflictive; this would require the existence of edges (f_r^i, f_s^j) and $(f_{r'}^i, f_s^j)$, what is impossible (Assumption 2). A set with 2 features cannot be conflictive (Assumptions 1 and 2), and a set with a single feature cannot be inconsistent by definition. Therefore, there will be no remaining inconsistencies or conflictive sets.   □

The main interest of the presented resolution algorithm is that it is fully distributed and it works on local information. Each robot uses its own $X_{ij}(t_i)$ for detecting the root robot of each conflictive set. During the resolution algorithm, the decisions, and actions taken by each robot are based on its local associations $\mathbf{A}_{ij}$, and the components assigned to its local features. Moreover, each robot is responsible of deleting the edges from its local association matrices $\mathbf{A}_{ij}$, with $j \in \{1, \ldots, n\}$. In addition, the presented algorithm works in finite time. Let us note that although we presented the algorithm for a single conflictive set, all conflictive sets are managed in parallel.

2.5 Feature Labeling

Simultaneously to the data association process, the robots assign labels to their features. After checking feature f_r^i is consistent, robot i assigns it a label $L_r^i = (i_\star, r_\star) \in \mathbb{N}^2$ composed of a robot identifier $i_\star$ and a feature index $r_\star$ as follows [2]. Assume f_r^i and features $f_s^j, f_{s'}^{j'}, \ldots$ form a consistent association set in $\mathscr{G}_{dis}$, and thus, they are observations of a common landmark in the environment taken by robots $i, j, j', \ldots$. Among all the candidates $(i, r), (j, s), (j', s'), \ldots$, a unique label $(i_\star, r_\star)$ is selected by the robots, e.g., the one with the lowest robot id. Then, robot i assigns this label to $f_r^i, L_r^i = (i_\star, r_\star)$; the other robots $j, j', \ldots$, proceed in a similar way so that finally,

$$L_r^i = L_s^j = L_{s'}^{j'} = \cdots = (i_\star, r_\star).$$

We say a feature f_r^i is *exclusive* if it is isolated in $\mathscr{G}_{dis}$, corresponding to a landmark observed by a single robot i; in this case, its label L_r^i is simply (i, r). Otherwise, we say f_r^i is nonexclusive and it may either be *consistent* or conflictive. Consistent features are labeled as explained above, whereas robots wait until conflicts are *resolved* for labeling its conflictive features. The data association and labeling process finishes with an association graph $\mathscr{G}_{dis}$ free of any inconsistent association and with all the features labeled. When the algorithm finishes, two features f_r^i, f_s^j have the same label, $L_r^i = L_s^j$, iff they are connected by a path in the resulting conflict-free $\mathscr{G}_{dis}$. The distributed data association and labeling algorithm is summarized in Algorithm 2.5.1. This strategy makes use of two subroutines to detect features and resolve inconsistencies that we explained in the previous sections.

Throughout this section, we use $\tilde{\mathscr{S}}_i \subseteq \mathscr{S}_i$ for the set of unlabeled features at robot $i \in \{1, \ldots, n\}$ and let $|\tilde{\mathscr{S}}_i|$ be its cardinality, i.e., the number of unlabeled features at robot i. The set of labels $\mathscr{L}_i$ consists of the labels L_r^i already assigned to the features $f_r^i \in \mathscr{S}_i \setminus \tilde{\mathscr{S}}_i$. Given a matrix X_{ij} of size $|\tilde{\mathscr{S}}_i| \times |\tilde{\mathscr{S}}_j|$, we define the function $\bar{r} = \text{row}\left(f_r^i\right)$ that takes an unlabeled feature $f_r^i \in \tilde{\mathscr{S}}_i$ and returns its associated row in X_{ij}, with $\bar{r} \in \{1, \ldots, |\tilde{\mathscr{S}}_i|\}$. Equivalently, we define the function $\bar{s} = \text{col}(f_s^j)$ for features in $\tilde{\mathscr{S}}_j$. We let $\tilde{\mathbf{A}}_{ij} \in \mathbb{N}^{|\tilde{\mathscr{S}}_i| \times |\tilde{\mathscr{S}}_j|}$ be like the local association matrix $\mathbf{A}_{ij}$, but containing exclusively the rows and columns of the unlabeled features of robots i and j.

Initially, all the features of each robot i are unlabeled,

$$\tilde{\mathscr{S}}_i = \{f_1^i, \ldots, f_{m_i}^i\}, \qquad \mathscr{L}_i = \emptyset.$$

Each robot i solves a local data association with each of its neighbors $j \in \mathscr{N}_i$ and obtains the association matrix $\mathbf{A}_{ij} \in \mathbb{N}^{m_i \times m_j}$. Then, the robot locally detects its *exclusive* features f_r^i which have not been associated to any other feature,

$$[\mathbf{A}_{ij}]_{r,s} = 0 \qquad \text{for all } j \in \mathscr{N}_i, \ j \neq i, \text{ and all } s \in \{1, \ldots, m_j\}. \tag{2.9}$$

Algorithm 2.5.1 Data association and labeling - Robot i

1: $\tilde{\mathscr{I}}_i \leftarrow \{f_1^i, \ldots, f_{m_i}^i\}$, $\mathscr{L}_i \leftarrow \emptyset$
2: Solve the local data association
3: ASSIGN_LABEL($L_r^i = (i, r)$, f_r^i) to each *exclusive* feature f_r^i
4: **while** $|\tilde{\mathscr{I}}_i| > 0$ **do**
5:     Run the detection algorithm 2
6:     Find each *consistent* feature f_r^i and its root $f_{r_\star}^{i_\star}$
7:     ASSIGN_LABEL($L_r^i = (i_\star, r_\star)$, f_r^i)
8:     Run the resolution algorithm
9:     Find each *resolved* feature f_r^i and its component id $[i_\star, r_\star]$
10:      ASSIGN_LABEL($L_r^i = (i_\star, r_\star)$, f_r^i)
11:      Find each *exclusive* feature f_r^i
12:      ASSIGN_LABEL($L_r^i = (i, r)$, f_r^i)
13: **end while**
14: **function** ASSIGN_LABEL(L_r^i, f_r^i)
15:     $\mathscr{L}_i \leftarrow \mathscr{L}_i \cup \{L_r^i\}$, $\tilde{\mathscr{I}}_i \leftarrow \tilde{\mathscr{I}}_i \setminus \{f_r^i\}$
16: **end function**

Since an exclusive feature f_r^i is always consistent, robot i assigns a label L_r^i to it, composed of its own robot id and feature index and removes it from the set of unlabeled features,

$$L_r^i = (i, r), \qquad \mathscr{L}_i = \mathscr{L}_i \cup L_r^i, \qquad \tilde{\mathscr{I}}_i = \tilde{\mathscr{I}}_i \setminus \{f_r^i\}. \qquad (2.10)$$

Since its unlabeled features in $\tilde{\mathscr{I}}_i$ may be conflictive, it executes the detection algorithm (2.4) on this subset.

The detection algorithm (Algorithm 2) is executed on the subgraph of $\mathscr{G}_{dis}$ involving the features in $\tilde{\mathscr{I}}_i$, for $i \in \{1, \ldots, n\}$. When it finishes, robot i has the power matrices $X_{ij} \in \mathbb{N}^{|\tilde{\mathscr{I}}_i| \times |\tilde{\mathscr{I}}_j|}$, for $j = 1, \ldots, n$, which contain the entries in $\mathbf{A}^{\text{diam}(\mathscr{G}_{dis})}$ associated to the features in $\tilde{\mathscr{I}}_i$ and $\tilde{\mathscr{I}}_j$. There is a path between f_r^i and f_s^j iff

$$[X_{ij}]_{\bar{r}, \bar{s}} > 0, \qquad (2.11)$$

being $\bar{r} = \text{row}(f_r^i)$ and $\bar{s} = \text{col}(f_s^j)$. These matrices give robot i the information about all the association paths of its features and the features of the rest of the robots in the network.

Then, each robot i detects its *consistent* features. After a feature f_r^i has been classified as consistent, its robot i proceeds to assign it a label. Here, we show how robot i decides the feature label $(i_\star, r_\star)$. Let us first give a general definition of the root robot of an either consistent or conflictive association set.

Definition 4 The *root* robot $i_\star$ for an association set is the one that has the most features in it. In case there are multiple candidates, it is the one with the lowest identifier. Equivalently, we define the *root* features $f_{r_\star}^{i_\star}, f_{r_\star'}^{i_\star}, \ldots$ as the features from the root robot that belong to the association set.   □

Using the power matrices $X_{i1}, \ldots, X_{in}$, robot i can find the number of features $\tilde{m}_j$ from a second robot j that belong to the same association set than f_r^i with $\bar{r} = \text{row}(f_r^i)$ as follows,

$$\tilde{m}_j = \left\| \left\{ f_s^j \mid [X_{ij}]_{\bar{r},\bar{s}} > 0, \text{ with } \bar{s} = \text{col}(f_s^j) \right\} \right\|. \tag{2.12}$$

If we let $\tilde{m}_\star$ be the maximum $\tilde{m}_j$ for $j \in \{1, \ldots, n\}$, then the root robot $i_\star$ and root features $f_{r_\star}^{i_\star}, f_{r'_\star}^{i_\star}, \ldots$ for the association set of f_r^i with $\bar{r} = \text{row}(f_r^i)$ are

$$i_\star = \min\left\{ j \mid \tilde{m}_j = \tilde{m}_\star \right\}, \quad \{r_\star, r'_\star, \ldots\} = \left\{ s \mid [X_{ii_\star}]_{\bar{r},\bar{s}} > 0 \text{ with } \bar{s} = \text{col}(f_s^{i_\star}) \right\}. \tag{2.13}$$

When f_r^i belongs to a consistent set, the root $i_\star$ corresponds to the robot with a single feature $f_{r_\star}^{i_\star}$ in the association set that has the lowest identifier,

$$\begin{aligned} i_\star &= \min\left\{ j \mid [X_{ij}]_{\bar{r},\bar{s}} > 0 \text{ for some } \bar{s} \in \{1, \ldots, |\tilde{\mathscr{S}}_j|\} \right\} \\ r_\star &= \left\{ s \mid [X_{ii_\star}]_{\bar{r},\bar{s}} > 0 \text{ with } \bar{s} = \text{col}(f_s^{i_\star}) \right\}, \end{aligned} \tag{2.14}$$

where $\bar{r} = \text{row}(f_r^i)$. Robot i assigns to its feature f_r^i the label $L_r^i = (i_\star, r_\star)$ and removes it from the set of unlabeled features,

$$L_r^i = (i_\star, r_\star), \qquad \mathscr{L}_i = \mathscr{L}_i \cup L_r^i, \qquad \tilde{\mathscr{S}}_i = \tilde{\mathscr{S}}_i \setminus \{f_r^i\}. \tag{2.15}$$

Thus, all features in the association set are assigned the same label. The robots proceed with all its consistent features in a similar fashion. For the features classified as conflictive, the resolution method (Algorithm 2.4.1) presented in the previous section is executed to solve the inconsistencies.

Let each component $\mathscr{C}_q$ in Algorithm 2.4.1 have the identifier $(i_\star, r_\star)$ composed of the root robot $i_\star$ and root feature $r_\star$ responsible of creating the component. When the resolution algorithm finishes, each feature f_r^i that has been assigned to a component $(i_\star, r_\star)$ has become consistent due to the edge removals. We say that such features are *resolved*. Thus, all the resolved features with the same component id form a consistent association set. Each robot i uses the component id of f_r^i as its label,

$$L_r^i = (i_\star, r_\star), \qquad \mathscr{L}_i = \mathscr{L}_i \cup L_r^i, \qquad \tilde{\mathscr{S}}_i = \tilde{\mathscr{S}}_i \setminus \{f_r^i\}. \tag{2.16}$$

Additionally, due to edge removal, some unlabeled features $f_r^i \in \tilde{\mathscr{S}}_i$ may have become exclusive. Robot i detects such features f_r^i by checking that

$$[\tilde{\mathbf{A}}_{ij}]_{\bar{r},\bar{s}} = 0, \quad \text{for all } j \in \mathscr{N}_i, \ j \neq i, \text{ all } \bar{s} \in \{1, \ldots, |\tilde{\mathscr{S}}_j|\},$$

being $\bar{r} = \text{row}(f_r^i)$, and it manages them as in (2.10). The remaining features may still be conflictive. Each robot i executes a new detection-resolution iteration on these still unlabeled features $\tilde{\mathcal{S}}_i$.

In a finite number of iterations, all features of all robots have been labeled, and the algorithm finishes. The interest of the presented algorithm is that it is fully distributed and works on local information. Each robot i uses its own X_{ij} to classify its features.

2.6 Algorithm Based on the Maximum Error Cut

The previous resolution algorithm has the advantage of solving all the inconsistencies in an easy way. However, the algorithm does not use information about the quality of the matches. When this information is available, it can be used to select which links should be broken to get rid of the inconsistent associations.

Most of the matching functions in the literature are based on errors between the matched features. These errors can be used to find a better partition of $\mathscr{C}$. Let $\mathbf{E}$ be the weighted symmetric association matrix

$$[\mathbf{E}]_{r,s} = \begin{cases} e_{rs} & \text{if } [\mathbf{A}]_{r,s} = 1, \\ -1 & \text{otherwise,} \end{cases} \tag{2.17}$$

with e_{rs} the error of the match between f_r and f_s.

Assumption 4 The error between matches satisfies:

- $e_{rr} = 0, \forall r$;
- Errors are nonnegative, $e_{rs} \geq 0, \forall r, s$;
- Errors are symmetric, $e_{rs} = e_{sr}, \forall r, s$;
- Errors of different matches are different, $e_{rs} = e_{r's'} \Leftrightarrow [r = r' \wedge s = s'] \vee [r = s' \wedge s = r']$;

□

Since the inconsistency is already known there is no need to use the whole matrix but just the submatrix related with the inconsistency, $\mathbf{E}_{\mathscr{C}}$. Although all the errors in $\mathbf{E}_{\mathscr{C}}$ are small enough to pass the matching between pairs of images, we can assume that the largest error in the path between two conflictive features is, with most probability, related to the spurious match.

Definition 5 Given two conflictive features, we define a bridge as a *single link* whose deletion makes the conflict between those two features disappear.   □

Note that not all the links in one inconsistency are bridges. There are links that, if deleted, would not break the inconsistency because:

- They do not belong to the path between the features to separate;
- They belong to the path, but they also belong to a cycle in the association graph, and therefore, they are not bridges.

Our goal is, for each pair of conflictive features, find and delete the bridge that links
them with the maximum error.

Algorithm 2.6.1 shows a solution to find the bridges using local interactions.
Along the section we explain in detail how it works. As we did in the detection algorithm (2.4), let each robot initialize its own rows of elements as $\mathbf{z}_r(0) =$
$\{[\mathbf{E}_{\mathscr{C}}]_{r,1}, \dots, [\mathbf{E}_{\mathscr{C}}]_{r,c}\}$, $r \in \{1, \dots, \tilde{m}_i\}$. Each robot manages the $\tilde{m}_i$ rows corresponding to the conflictive features it has observed. The update rule executed by
every robot and every feature is

$$\mathbf{z}_r(t+1) = \max_{s \in \mathscr{C}, \, [\mathbf{E}_{\mathscr{C}}]_{r,s} \geq 0} (\mathbf{z}_r(t), \mathbf{z}_s(t)\mathbf{P}_{rs}), \qquad (2.18)$$

where the maximum is done element to element and $\mathbf{P}_{rs}$ is the permutation matrix
of the columns r and s. We have dropped the super indices corresponding to robots
because the limited communications are implicit in the error caused by direct associations, Eq. (2.17).

Algorithm 2.6.1 *Maximum Error Cut* - Robot i

Require: Set of $\mathscr{C}$ different conflictive sets
Ensure: $\mathscr{G}_{dis}$ is *conflict free*
1: **for all** $\mathscr{C}$ **do**
2:      – *Error transmission*
3:      $\mathbf{z}_r(0) = \{[\mathbf{E}_{\mathscr{C}}]_{r,1}, \dots, [\mathbf{E}_{\mathscr{C}}]_{r,c}\}$, $r = 1, \dots, \tilde{m}_i$
4:      **repeat**
5:          $\mathbf{z}_r(t+1) = \max_{s \in \mathscr{C}, \, [\mathbf{E}_{\mathscr{C}}]_{r,s} \geq 0}(\mathbf{z}_r(t), \mathbf{z}_s(t)\mathbf{P}_{rs})$
6:      **until** $\mathbf{z}_r(t+1) = \mathbf{z}_r(t)$, $\forall r \in \tilde{m}_i$
7:      – *Link Deletion*
8:      **while** robot i has conflictive features r and r' **do**
9:          Find the bridges (s, s') :
10:             (a) $[\mathbf{z}_r]_s = [\mathbf{z}_{r'}]_{s'}$, $s \neq s'$,
11:             (b) For all $s'' \neq s$, $[\mathbf{z}_r]_s \neq [\mathbf{z}_r]_{s''}$,
12:             (c) For all $s'' \neq s'$, $[\mathbf{z}_{r'}]_{s'} \neq [\mathbf{z}_{r'}]_{s''}$
13:          Select the bridge with largest error
14:          Send message to break it
15:      **end while**
16: **end for**

Proposition 4 *The dynamic system defined in* (2.18) *converges in a finite number
of iterations and for any* $r, s \in \mathscr{C}$ *such that* $[\mathbf{E}_{\mathscr{C}}]_{r,s} \geq 0$ *the final value of* $\mathbf{z}_r$ *is the
same than* $\mathbf{z}_s\mathbf{P}_{rs}$.

Proof The features involved in the inconsistency form a strongly connected graph.
For a given graph, the max consensus update is proved to converge in a finite num-
ber of iterations [6]. For any $r, s \in \mathscr{C}$ such that $[\mathbf{E}_{\mathscr{C}}]_{r,s} \geq 0$, by Eq. (2.18) and the
symmetry of $\mathbf{E}_{\mathscr{C}}$, the final consensus values of $\mathbf{z}_r$ and $\mathbf{z}_s$ satisfy, element to element
that

$$\mathbf{z}_r \geq \mathbf{z}_s \mathbf{P}_{rs} \text{ and } \mathbf{z}_s \geq \mathbf{z}_r \mathbf{P}_{sr} \tag{2.19}$$

Using the properties of the permutation matrices, $\mathbf{P}_{rs} = \mathbf{P}_{sr} = \mathbf{P}_{sr}^{-1}$, we see that $\mathbf{z}_s \mathbf{P}_{rs} \geq \mathbf{z}_r$, which combined with Eq. (2.19) yields to $\mathbf{z}_r = \mathbf{z}_s \mathbf{P}_{rs}$.   □

Let us see the convergence values of the different elements. Considering again Eq. (2.18) for a given feature f_r, we can express it as a function of its elements and the uth component, $[\mathbf{z}_r(t+1)]_u$, is updated as follows:

$$[\mathbf{z}_r(t+1)]_u$$
$$= \begin{cases} \max([\mathbf{z}_r(t)]_u, [\mathbf{z}_s(t)]_s) & \text{if } [\mathbf{E}_{\mathscr{C}}]_{r,s} \geq 0 \wedge u = r \\ \max([\mathbf{z}_r(t)]_u, [\mathbf{z}_s(t)]_r) & \text{if } [\mathbf{E}_{\mathscr{C}}]_{r,s} \geq 0 \wedge u = s \\ \max([\mathbf{z}_r(t)]_u, [\mathbf{z}_s(t)]_u) & \text{if } [\mathbf{E}_{\mathscr{C}}]_{r,s} \geq 0 \wedge r \neq u \neq s \end{cases}, \tag{2.20}$$

where the two first rows are due to the permutations. Let us first analyze the case in which the inconsistency does not contain any cycle.

Theorem 3 *If $\mathscr{C}$ is cycle free, then:*

(i) For any $r \in \mathscr{C}$, $[\mathbf{z}_r(t)]_r = 0, \forall t \geq 0$.
(ii) $[\mathbf{z}_r(t)]_{s'} \to [\mathbf{E}_{\mathscr{C}}]_{r',s'} = e_{r's'}$, where

$$r' = \underset{[A]_{r'',s'}=1}{\arg \min}\, d(r, r''),$$

and $d(r, r'')$ is the distance in links to reach node r'' starting from node r. In other words, $f_{r'}$ is the closest feature to f_r directly associated to $f_{s'}$.

Proof For any feature, f_r, taking into account Eq. (2.20), the update of the rth element of $\mathbf{z}_r$, $[\mathbf{z}_r(t+1)]_r$, is

$$[\mathbf{z}_r(t+1)]_r = \underset{s \in \mathscr{C}, [\mathbf{E}_{\mathscr{C}}]_{r,s} \geq 0}{\max} ([\mathbf{z}_r(t)]_r, [\mathbf{z}_s(t)]_s).$$

Recalling the first point in Assumption 4, the initial value of $[\mathbf{z}_r(0)]_r = e_{rr} = 0$, for all r, then $[\mathbf{z}_r(t)]_r = 0, \forall t \geq 0$.

The inconsistency does not have any cycles and there is a path between any two features, the conflict is a spanning tree. Let us consider one link, $(f_{r'}, f_{s'})$. The link creates a partition of $\mathscr{C}$ in two strongly connected, disjoint subsets

$$\mathscr{C}_{r'} = \{r \mid d(r, r') < d(r, s')\},$$

$$\mathscr{C}_{s'} = \{s \mid d(s, s') < d(s, r')\}.$$

In the above equations it is clear that $r' \in \mathscr{C}_{r'}$ and $s' \in \mathscr{C}_{s'}$.

We will focus now on the values of the s'th element of the state vector for the nodes in $\mathscr{C}_{r'}$ and the r'th element for the nodes in $\mathscr{C}_{s'}$,

$$[\mathbf{z}_r(t)]_{s'}, r \in \mathscr{C}_{r'}, \text{ and } [\mathbf{z}_s(t)]_{r'}, s \in \mathscr{C}_{s'}.$$

In the first case, for any $r \in \mathscr{C}_r \backslash r'$, update rule (2.20) is equal to

$$[\mathbf{z}_r(t+1)]_{s'} = \max_{r'' \in \mathscr{C}_{r'}, [\mathbf{E}_{\mathscr{C}}]_{r,r''} \geq 0} ([\mathbf{z}_r(t)]_{s'}, [\mathbf{z}_{r''}(t)]_{s'}),$$

because $r \neq s' \neq r''$. The nodes in $\mathscr{C}_{s'}$ are not taken into account because that would mean that $\mathscr{C}$ has a cycle. The special case of feature $f_{r'}$ has an update rule equal to

$$[\mathbf{z}_{r'}(t+1)]_{s'} = \max_{r \in \mathscr{C}_{r'}, [\mathbf{E}_{\mathscr{C}}]_{r',r} \geq 0} ([\mathbf{z}_{r'}(t)]_{s'}, [\mathbf{z}_r(t)]_{s'}, [\mathbf{z}_{s'}(t)]_{r'}).$$

In a similar way the updates for features in $\mathscr{C}_s$ are

$$[\mathbf{z}_s(t+1)]_{r'} = \max_{s'' \in \mathscr{C}_{s'}, [\mathbf{E}_{\mathscr{C}}]_{s,s''} \geq 0} ([\mathbf{z}_s(t)]_{r'}, [\mathbf{z}_{s''}(t)]_{r'}),$$

$$[\mathbf{z}_{s'}(t+1)]_{r'} = \max_{s \in \mathscr{C}_{s'}, [\mathbf{E}_{\mathscr{C}}]_{s',s} \geq 0} ([\mathbf{z}_{s'}(t)]_{r'}, [\mathbf{z}_s(t)]_{r'}, [\mathbf{z}_{r'}(t)]_{s'}).$$

Considering together all the equations and the connectedness of $\mathscr{C}_{r'}$ and $\mathscr{C}_{s'}$, all these elements form a connected component and they will converge to

$$\max_{r \in \mathscr{C}_{r'}, s \in \mathscr{C}_{s'}} ([\mathbf{z}_r(0)]_{s'}, [\mathbf{z}_s(0)]_{r'}).$$

Since all the features $r \in \mathscr{C}_{r'} \backslash r'$ are not associated with $f_{s'}$, $[\mathbf{z}_r(0)]_{s'} = -1$. Analogously, for all the features $s \in \mathscr{C}_{s'} \backslash s'$, $[\mathbf{z}_s(0)]_{r'} = -1$. Finally, for the features r' and s', by the second and third point of Assumption 4, $[\mathbf{z}_{r'}(0)]_{s'} = e_{r's'} = e_{s'r'} = [\mathbf{z}_{s'}(0)]_{r'} \geq 0 > -1$. Therefore, this subset of c elements of the state vectors converge to the error of the link $(f_{r'}, f_{s'})$, $e_{r's'}$. From Proposition 4 we can also see that for any $r \in \mathscr{C}_{r'}$, $[\mathbf{z}_r]_s$, $s \in \mathscr{C}_{s'} \backslash s'$, will converge to the final value of $[\mathbf{z}_{s'}]_s$.

The same argument applies for the rest of the links and the proof is complete. □

Let us see what happens now in the presence of cycles in the inconsistency.

Theorem 4 *Let us suppose the inconsistency has a cycle involving ℓ features. Let $\mathscr{C}_\ell$ be the subset of features that belong to the cycle. After the execution of (2.18) it holds that:*

(i) $\forall r', s' \in \mathscr{C}_\ell, s' \neq r'$

$$[z_{r'}]_{s'} \to \max_{r,s \in \mathscr{C}_\ell} e_{rs}.$$

(ii) $\forall r' \notin \mathscr{C}_\ell, s' \in \mathscr{C}_\ell, s' \neq \arg\min_{s \in \mathscr{C}_\ell} d(r', s),$

$$[z_{r'}]_{s'} \to \max_{r,s \in \mathscr{C}_\ell} e_{rs}.$$

Proof In the proof, we will denote $r_1, \ldots, r_\ell$, the set of features in $\mathscr{C}_\ell$. Without loss of generality we will assume that the links that form the cycle are (f_{r_1}, f_{r_2}), $(f_{r_2}, f_{r_3}), \ldots, (f_{r_\ell}, f_{r_1})$. For an easy reading of the proof of this result we will omit the time indices in the update equations. Let us consider the update rule (2.20) for element r_2 of feature f_{r_1},

$$[\mathbf{z}_{r_1}]_{r_2} = \max([\mathbf{z}_{r_1}]_{r_2}, [\mathbf{z}_{r_2}]_{r_1}, [\mathbf{z}_{r_\ell}]_{r_2}),$$

where we have also omitted other possible features that are directly linked to f_{r_1} because if they are also linked to f_{r2} they belong to $\mathscr{C}_\ell$ and if not they do not affect to the final result.

From the above equation we observe that $[\mathbf{z}_{r_1}]_{r_2}$ depends on the value of $[\mathbf{z}_{r_2}]_{r_1}$. At the same time this value is updated with

$$[\mathbf{z}_{r_2}]_{r_1} = \max([\mathbf{z}_{r_2}]_{r_1}, [\mathbf{z}_{r_1}]_{r_2}, [\mathbf{z}_{r_3}]_{r_1}),$$

which depends on the value of $[\mathbf{z}_{r_3}]_{r_1}$. If we keep with the chain of associations we reach the point in which $[\mathbf{z}_{r_{\ell-1}}]_{r_1}$ depends on $[\mathbf{z}_{r_\ell}]_{r_1}$, which has update rule equal to

$$[\mathbf{z}_{r_\ell}]_{r_1} = \max([\mathbf{z}_{r_\ell}]_{r_1}, [\mathbf{z}_{r_{\ell-1}}]_{r_1}, [\mathbf{z}_{r_1}]_{r_\ell}).$$

As we have proved in Proposition 4, in the end $[\mathbf{z}_{r_1}]_{r_2} = [\mathbf{z}_{r_2}]_{r_1}$, $[\mathbf{z}_{r_2}]_{r_1} = [\mathbf{z}_{r_3}]_{r_1}, \ldots, [\mathbf{z}_{r_{\ell-1}}]_{r_1} = [\mathbf{z}_{r_\ell}]_{r_1}$ and $[\mathbf{z}_{r_\ell}]_{r_1} = [\mathbf{z}_{r_1}]_{r_\ell}$ because they are direct neighbors. This means that after the execution of enough iterations of (2.18), $[\mathbf{z}_{r_1}]_{r_2} = [\mathbf{z}_{r_1}]_{r_\ell} = [\mathbf{z}_{r_1}]_{r_1}, \forall r \in \mathscr{C}_\ell \backslash r_1$. By applying the same argument for any other feature in $\mathscr{C}_\ell$ we conclude that after the execution of the update, for any $r \in \mathscr{C}_\ell$, $[\mathbf{z}_r]_{r'} = [\mathbf{z}_r]_{r''}, \forall r', r'' \in \mathscr{C}_\ell \backslash r$. Thus, each feature inside the cycle will end with $\ell - 1$ elements in its state vector with the same value (the maximum of all the considered links) and (*i*) is true. If there are any additional links inside the cycle the result is the same including in the max consensus the weights of these links.

Now let us consider the rest of the features in the inconsistency, $\bar{\mathscr{C}}_\ell = \mathscr{C} \backslash \mathscr{C}_\ell$. Given a feature $s \in \bar{\mathscr{C}}_\ell$ two things can happen:

- ∃ unique $r \in \mathscr{C}_\ell$ such that f_r and f_s are directly associated;
- s is not directly associated with any feature in $\mathscr{C}_\ell$ but there exists at least one path of features $\in \bar{\mathscr{C}}_\ell$ that ends in a unique feature $r \in \mathscr{C}_\ell$.

The uniqueness of r comes from the fact that if there were another feature $r' \in \mathscr{C}_\ell$, reachable from s without passing through r, that would mean that s is also part of the cycle. Note that this does not discard the possibility that r and s belong to another cycle different than $\mathscr{C}_\ell$.

As we have seen in the proof of Theorem 3, due to the fact that r is the only connection with $\mathscr{C}_\ell$, for any $r' \in \mathscr{C}_\ell \backslash r$, $[\mathbf{z}_s]_{r'}$ will have final value equal to $[\mathbf{z}_r]_{r'}$ which proves (*ii*). On the other hand, $[\mathbf{z}_s]_r$ will have the value of the link that connects it to feature r or, if f_r and f_s belong to another cycle different than $\mathscr{C}_\ell$, the maximum error of all the links that form the second cycle. In both cases, doing a change in the names of the indices, we can see that (*ii*) is also true. ☐

At this point we are ready to define the bridges in terms of the variables $\mathbf{z}_r$ and to propose a criterion to select the bridge to break. The bridges, $(f_s, f_{s'})$, for any pair of conflictive features f_r and $f_{r'}$ satisfy

(a) $[\mathbf{z}_r]_s = [\mathbf{z}_{r'}]_{s'}, s \neq s'$,
(b) for all $s'' \neq s$, $[\mathbf{z}_r]_s \neq [\mathbf{z}_r]_{s''}$,
(c) for all $s'' \neq s'$, $[\mathbf{z}_{r'}]_{s'} \neq [\mathbf{z}_{r'}]_{s''}$.

The first condition comes from Theorem 3 and the other two come from Theorem 4. Note that for any bridge, the error of the bridge is the same as the value of $[\mathbf{z}_r]_s$, $[\mathbf{z}_r]_s = [\mathbf{z}_{r'}]_{s'} = e_{ss'}$. Therefore, each node can look in a local way at its own rows and choose the best bridge that breaks the conflict, the one with the largest error. In case one robot has more than two features in the same conflict, finding the optimal cut becomes NP-hard. In this chapter, we use a greedy approach that returns good results. Our solution chooses two of the $\tilde{m}_i$ inconsistent features and selects the best bridge for them. The bridge separates all the $\tilde{m}_i$ features in two disconnected subsets. The process is repeated with each of the subsets until the inconsistencies are solved.

Note that we are considering only single-link deletions. Cycles in the association graph are sets of features strongly associated, and therefore, it is better not to break links there. If two conflictive features belong to the same cycle, then there are no bridges. However, the algorithm is also able to detect this situation and the *Spanning Trees* can be used to solve the conflict.

In conclusion, this algorithm is able to detect in a local way the best bridge to break each inconsistency. This provides a more solid criterion to solve the inconsistencies than just cutting arbitrary edges. Each robot is able to detect which set of links is best to cut in order to solve the conflicts regarding its own features. The algorithm also finishes in finite time and does not require much additional bandwidth because, as in the detection algorithm, the amount of transmitted information can be optimized. An example of execution of the algorithm is given in Fig. 2.3 and more details can be found in [18].

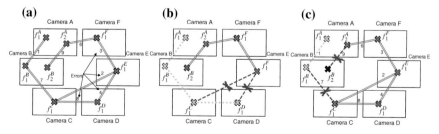

Fig. 2.3 Example of execution of the resolution of one inconsistency using the two approaches. **a** Inconsistency. **b** Solution obtained using the *Spanning Trees* algorithm. **c** Solution obtained using the *Maximum Error Cut* approach

2.7 Simulations

We have carried out several simulations with a team composed by 7 robots exploring an environment of 20×20 m with 300 features (Fig. 2.4). Each robot executes 70 motion steps along a path of approximately 30 m. The robots estimate their motion based on odometry information that is corrupted with a noise of standard deviation $\sigma_x, \sigma_y = 0.4$ cm for the translations and $\sigma_\theta = 1$ degree for the orientations. They sense the environment using an omnidirectional camera that gives bearing measurement to features within 360 degrees around the robot and within a distance of 6 m. The measurements are corrupted with a noise of 0.5 degrees standard deviation. Each robot explores the environment and builds its local map (Fig. 2.4b). Due to the presence of obstacles (gray areas), each robot may have not observed some landmarks.

When they finish the exploration, they execute the distributed data association algorithm explained in this chapter under the communication graph in Fig. 2.5a. The local data associations $F(\mathscr{S}_i, \mathscr{S}_j)$ are obtained by applying the JCBB method [19] to the local maps of any pair of neighboring robots $(i, j) \in \mathscr{E}_{com}$. Since all the trajectories followed by the robots traverse the main corridor (Fig. 2.4) there is a high overlapping between their local maps (Table 2.1). Given any 2 local maps with approx. 122 features, there are approximately 89 true matches (ground truth). Although, the local data association method has found a high amount of the ground truth links (good links or true positives), it has also missed a few of them (missing links or false negatives). In addition, some additional links have been detected that link together different features (spurious links or false positives). From the 858 features within all the local maps, there are 300 different features in the ground truth sense (association sets). From them, 184 were observed by a single robot (ground truth exclusive features), and the remaining where observed by around 6 robots (ground truth size of the

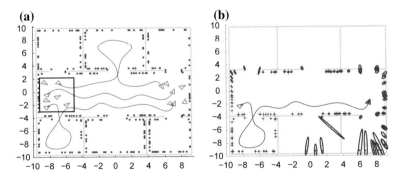

Fig. 2.4 A team of 7 robots explore an environment of 20×20 m. **a** *Gray areas* are walls and *red dots* are the ground truth location of landmarks. The robots (*triangles*) start in the left (*black box region*) and finish in the right. **b** Local map estimated by robot 2. The landmarks close to its trajectory (*red line*) have been estimated (*blue crosses* and *ellipses*) with a high precision. Due to the presence of obstacles (*gray areas*) some of the landmarks have not been observed, or have been estimated with high uncertainty

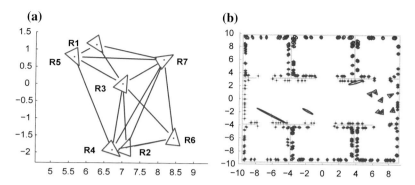

Fig. 2.5 **a** Communication graph associated to the final robot poses in Fig. 2.4. There is a link (*blue solid line*) between any pair of robot poses (*red triangles*) that are within a distance of 3 m. **b** Global map obtained after merging the local maps. *Red dots* and *triangles* are the ground truth position of the features and robot poses. The estimated feature positions are shown with blue crosses and ellipses. The map merging process is explained in detail in Chap. 4; here we display the global map estimated by robot 2 after $t = 5$ iterations of the map merging algorithm

Table 2.1 Local data associations

Features	Per local map	Total
Features observed	122	**858**
Data associations	Per pair of local maps	Total
Links (ground truth)	89	2860
Links	88	2820
Good links	85	2750
Missing links	3	110
Spurious links	2	70
Association sets	Obtained	Ground truth
Association sets	**296**	**300**
Exclusive features	**187**	**184**
Nonexclusive assoc.	**109**	116
Size of nonexclusive	6.1	5.8

nonexclusive). In the data association graph $\mathcal{G}_{dis}$ however, only 296 association sets have been obtained, which means that different features have been mixed up together. There are 184 exclusive features (ground truth exclusive features), although the local data association algorithm has found 187 exclusive features. These additional three exclusive features appear due to the presence of the three outliers, the features with high covariance ellipses in Fig. 2.5b. Since their positions have been wrongly estimated, the local data association method has failed to correctly associate them.

The robots execute Algorithm 2.5.1 on the nonexclusive features to propagate the local matches and discover the associations between their features and the ones

Table 2.2 Detection and resolution of inconsistent associations

Detection	Conflictive	Consistent nonexclusive	Consistent exclusive
Association sets	7	102	187
Features	80	591	187
Resolution	Conflictive	Consistent nonexclusive	Consistent exclusive
Association sets	0	116 (**+14**)	187
Features	0	671 (**+80**)	187

observed by the other team members. In addition, they establish the labels for their features, and they detect and solve any inconsistent associations. From the 109 nonexclusive association sets, 102 of them are consistent, and its associated 591 features are classified as consistent (Table 2.2). The remaining seven sets are conflictive, and they have associated 80 conflictive features. After executing the resolution algorithm on the 80 conflictive features, all of them are resolved and the process finishes. The original seven conflictive sets are partitioned into 14 consistent nonexclusive sets. Due to these additional sets, the number of consistent nonexclusive association sets (Table 2.2, third row), which initially was 102 (Table 2.2, first row), is increased into 116 (102 + 14) after executing the algorithm. Equivalently, the number of consistent nonexclusive features (Table 2.2, fourth row) which was 591 (Table 2.2, second row) becomes 671 (591 + 80) since the 80 inconsistent features are resolved.

Table 2.3 compares the final data association graph and the ground truth information. Since the resolution algorithm is based on link deletion, the number of links here is lower than in Table 2.1. However, the number of association sets is closer to the ground truth results. From the 303 obtained association sets, three of them

Table 2.3 Results after detecting and solving the inconsistencies

Features	Per local map	Total
Features observed	122	858
Data associations	Per pair of local maps	Total
Links (ground truth)	89	2860
Links	87	2794 (**−26**)
Good links	85	2746 (**−4**)
Missing links	3	114 (**+4**)
Spurious links	2	48 (**−22**)
Association sets	Obtained	Ground truth
Association sets	303	300
Exclusive features	187	184
Nonexclusive assoc.	116	116
Size of nonexclusive	5.7	5.8

are due to the three outliers in Fig. 2.5b. Thus, there are 300 remaining association sets, which is exactly the same number of association sets in the ground truth data. The same behavior is observed regarding their sizes. This means that the resulting associations are similar to the ground truth ones in spite of the fact that they have less links. From the 26 links erased from $\mathscr{G}_{dis}$, 22 were spurious links, and only 4 where good links that now are missing. Robots use the obtained data association for computing the global map (Fig. 2.5b) as described in Chap. 4.

2.8 Closure

In this chapter, we have presented a distributed technique to match sets of features observed by a team of robots in a consistent way under limited communications. Local associations are found only within robots that are neighbors in the communication graph. After that, a fully distributed method to compute all the paths between local associations is carried out, allowing the robots to detect all the inconsistencies related with their observations. For every conflictive set detected, in a second step the method is able to delete local associations to break the conflict using only local communications. The whole method is proved to finish in a finite amount of time finding and solving all the inconsistent associations. We have studied the performance of the method for robots equipped with omnidirectional cameras in a simulated environment. Additional experiments with real data acquired with RGB-D and conventional cameras are presented in Chap. 5.

References

1. R. Aragues, E. Montijano, C. Sagues, Consistent data association in multi-robot systems with limited communications, in *Robotics: Science and Systems*, Zaragoza, Spain, June 2010
2. R. Aragues, J. Cortes, C. Sagues, Distributed consensus algorithms for merging feature-based maps with limited communication. Robot. Auton. Syst. **59**(3–4), 163–180 (2011)
3. S. Avidan, Y. Moses, Y. Moses, Centralized and distributed multi-view correspondence. Int. J. Comput. Vis. **71**(1), 49–69 (2007)
4. T. Bailey, H. Durrant-Whyte, Simultaneous localization and mapping: part II. IEEE Robot. Autom. Mag. **13**(3), 108–117 (2006)
5. T. Bailey, E.M. Nebot, J.K. Rosenblatt, H. Durrant-Whyte, Data association for mobile robot navigation: a graph theoretic approach, in *IEEE International Conference on Robotics and Automation*, San Francisco, USA, April 2000, pp. 2512–2517
6. F. Bullo, J. Cortes, S. Martinez, *Distributed Control of Robotic Networks*. Applied Mathematics Series (Princeton University Press, Princeton, 2009), http://coordinationbook.info
7. C. Cadena, F. Ramos, J. Neira, Efficient large scale SLAM including data association using the Combined Filter, in *European Conference on Mobile Robotics*, Mlini/Dubrovnik, Croatia, September 2009, pp. 217–222
8. A. Censi, An accurate closed-form estimate of ICP's covariance, in *IEEE International Conference on Robotics and Automation*, Roma, Italy, April 2007, pp. 3167–3172
9. R.W. Deming, L.I. Perlovsky, Concurrent multi-target localization, data association, and navigation for a swarm of flying sensors. Inf. Fusion **8**(3), 316–330 (2007)

10. V. Ferrari, T. Tuytelaars, L. Van Gool, Wide-baseline multiple-view correspondences, in *IEEE International Conference on Computer Vision and Pattern Recognition*, Madison, USA, June 2003, pp. 718–725
11. M.A. Fischler, R.C. Bolles, Random sample consensus: a paradigm for model fitting with applications to image analysis and automated cartography. Commun. ACM **24**(6), 381–395 (1981)
12. D. Fox, J. Ko, K. Konolige, B. Limketkai, D. Schulz, B. Stewart, Distributed multirobot exploration and mapping. IEEE Proc. **94**(7), 1325–1339 (2006)
13. A. Gil, O. Reinoso, M. Ballesta, M. Julia, Multi-robot visual SLAM using a rao-blackwellized particle filter. Robot. Auton. Syst. **58**(1), 68–80 (2009)
14. A. Howard, Multi-robot simultaneous localization and mapping using particle filters. Int. J. Robot. Res. **25**(12), 1243–1256 (2006)
15. M. Kaess, F. Dellaert, Covariance recovery from a square root information matrix for data association. Robot. Auton. Syst. **57**(12), 1198–1210 (2009)
16. K. Konolige, J. Gutmann, B. Limketkai, Distributed map-making, in *Workshop on Reasoning with Uncertainty in Robotics, International Joint Conference on Artificial Intelligence*, Acapulco, Mexico, August 2003
17. H.S. Lee, K.M. Lee, Multi-robot SLAM using ceiling vision, in *IEEE/RSJ International Conference on Intelligent Robots and Systems*, St. Louis, USA, October 2009, pp. 912–917
18. E. Montijano, R. Aragues, C. Sagues, Distributed data association in robotic networks with cameras and limited communications. IEEE Trans. Robot. **29**(6), 1408–1423 (2013)
19. J. Neira, J.D. Tardós, Data association in stochastic mapping using the joint compatibility test. IEEE Trans. Robot. Autom. **17**(6), 890–897 (2001)
20. C.H. Papadimitriou, K. Steiglitz, The max-flow, min-cut theorem (chapter 6.1), *Combinatorial Optimization: Algorithms and Complexity* (Dover Publications, New York, 1998), pp. 120–128
21. M. Pfingsthorn, B. Slamet, A. Visser, A scalable hybrid multi-robot SLAM method for highly detailed maps, in *RoboCup 2007: Robot Soccer World Cup XI*, Lecture Notes in Artificial Intelligence, vol. 5001, ed. by U. Visser, F. Ribeiro, T. Ohashi, F. Dellaert (Springer, Berlin, 2008), pp. 457–464
22. S. Thrun, Y. Liu, Multi-robot SLAM with sparse extended information filters, in *International Symposium of Robotics Research*, Italy, Sienna, October 2003, pp. 254–266
23. S.B. Williams, H. Durrant-Whyte, Towards multi-vehicle simultaneous localisation and mapping, in *IEEE International Conference on Robotics and Automation*, Washington, DC, USA, May 2002, pp. 2743–2748
24. X.S. Zhou, S.I. Roumeliotis, Multi-robot SLAM with unknown initial correspondence: the robot rendezvous case, in *IEEE/RSJ International Conference on Intelligent Robots and Systems*, Beijing, China, October 2006, pp. 1785–1792

Chapter 3
Distributed Localization

Abstract In this chapter we study the problem of distributed localization, which consists of establishing a common frame and computing the robots' localization relative to this frame. Each robot is capable of measuring the relative pose of its neighboring robots. However, it does not know the poses of far robots, and it can only exchange data with neighbors using the range-limited communication network. The analyzed algorithms have the interesting property that can be executed in a distributed fashion. They allow each robot to recover localization using exclusively local information and local interactions with its neighbors. Besides, they only require each robot to maintain an estimate of its own pose. Thus, the memory load of the algorithm is low compared to methods where each robot must also estimate the poses of any other robot. We analyze two different scenarios and study distributed algorithms for them. In the first scenario each robot measures the noisy planar position and orientation of nearby robots to estimate its own full localization with respect to an anchor node. In the second case, robots take noisy measurements of the relative three-dimensional positions of their neighbors, which is used to estimate their three-dimensional positions with respect to the simultaneously computed centroid reference. When the centroid of the team is selected as common frame, the estimates are more precise than with any anchor selection.

Keywords Localization · Limited communication · Distributed systems · Parallel computation

3.1 Introduction

Multi-robot tasks, such as pattern formation [7, 36] or inter-robot collision avoidance [32], often require the knowledge of the robots' positions in a common reference frame. Typically, robots start at unknown locations, they do not share any common frame, and they can only measure the relative positions of nearby robots. We address the localization problem, which consists of combining these relative measurements to build an estimate of the robots' localization in a common frame.

Several localization algorithms rely on range-only [1, 10, 11], or bearing-only [31] relative measurements of the robots' poses. Other approaches assume that robots

© The Author(s) 2015
R. Aragues et al., *Parallel and Distributed Map Merging and Localization*,
SpringerBriefs in Computer Science, DOI 10.1007/978-3-319-25886-7_3

measure the full state of their nearby robots. The relative full-pose of a pair of robots can be obtained, for instance, by comparing their local maps [14, 15, 33] and looking for overlapping regions. This approach, known as map alignment, presents a high computational cost and its results depend on the accumulated uncertainty in the local maps. Alternatively, each robot can locally combine several observations to build an estimate of the relative poses. The 2D relative pose can be retrieved from at least five noisy distance measurements and four noisy displacements [38]. Bearing-only measurements can be also used to recover the 2D relative pose in vision systems [20]. The 3D case has also been analyzed for distance and bearing, bearing-only, and distance-only observations [35]. These methods present the benefit that the obtained results do not depend on the uncertainties in the local maps. They also allow the robots to compute their relative poses when there is no overlapping between their maps, or even if they do not actually have a map.

Localization algorithms in networked systems properly combine the previous relative measurements to produce an estimate of the robots' poses. Some distributed algorithms compute both, the positions and orientations but assume that the relative measurements are noise free, e.g., [19] where each robot reaches an agreement on the centroid of the network expressed in its local reference frame. Other methods compute exclusively the robot positions but not their orientations, and consider noisy relative measurements of the robot positions. This latter localization problem can be solved by using linear optimization methods [4, 28]. Although these works do not consider the robots' orientations, they can also be applied to such cases provided that the robots have previously executed an attitude synchronization [25, 30] or a motion coordination [16] strategy to align their orientations. Cooperative localization algorithms [22, 27, 34] do not just compute the network localization once, but also track the robots positions. These algorithms, however, usually assume that an initial guess on the robot poses exists.

Formation control [16, 18, 21, 24] and network localization are related problems. While localization algorithms compute robot positions that satisfy the inter-robot restrictions, in formation control problems the robots actually move to these positions. The goal formation is defined by a set of inter-robot restrictions (range-only, bearing-only, full-positions, or relative poses). Although some works discuss the effects of measurement noises in the final result [16], formation algorithms usually assume that both, the measurements and the inter-robot restrictions are noise free [18, 21, 24]. Thus, additional analysis is necessary in noisy localization scenarios.

Both, formation control and localization problems can be solved up to a rotation and a translation. This ambiguity disappears when the positions of a subset of *anchor* robots is given in some absolute reference frame. The range-only case [1] requires at least three non-collinear anchors for planar scenarios. The density and placement of anchors has an important effect on the accuracy of the solution for the bearing-only case [31]. In the full-position case a single anchor is enough. Its placement influences the accuracy of the final results and it is common to analyze the estimation errors at the robots as a function of their distances to the anchor [6]. However, it is common to assume that the first robot is the anchor placed at the origin of the common reference frame and make the other robots compute their positions relative to the anchor.

In this chapter we focus on network localization methods where robots measure the relative pose of their neighbors. Since these methods do not require the robots to have a map, they can be executed at any time. In particular, we execute it at an initial stage, prior to any exploration taking place. The communication graph during this initial stage must be connected. We consider scenarios with noisy relative measurements. We assume that these measurements are independent, since they are acquired individually by the robots. We do not further discuss cooperative localization algorithms, since in a map merging scenario it is enough for the robots to compute the global frame and their poses once. In addition, we discuss the selection of the common reference frame. We consider the cases that the common frame is one of the robots (anchor-based), and that the common frame is the centroid. Firstly we present a distributed algorithm for planar scenarios. Each agent uses noisy measurements of relative planar poses with respect to other robots to estimate its planar localization relative to an anchor node. After that, we discuss the localization problem for higher dimension scenarios. We present a distributed algorithm that allows each robot to simultaneously compute the centroid of the team and its positions relative to the centroid. We show that when the centroid of the team is selected as the common frame, the estimates are more precise than with any anchor selection.

In order to make the reading easy, along the chapter we use the indices i, j to refer to robots and indices e, e' to refer to edges. An edge e starting at robot i and ending at robot j is represented by $e = (i, j)$. Given a matrix A, the notations $A_{r,s}$ and $[A]_{r,s}$ corresponds to the (r, s) entry of the matrix. We let $\otimes$ be the Kronecker product, $\mathbf{I}_r$ be the identity matrix of size $r \times r$, and $\mathbf{0}_{r \times s}$ be a $r \times s$ matrix with all entries equal to zero. A matrix A defined by blocks A_{ij} is denoted $A = [A_{ij}]$. The operation $A = \text{blkDiag}(B_1, \ldots, B_r)$ returns a matrix A defined by blocks with $A_{ii} = B_i$ and $A_{ij} = \mathbf{0}$ for $i \neq j$.

3.2 Problem Description

The problem addressed in this chapter consists of computing the localization of a network of $n \in \mathbb{N}$ robots from relative measurements. We consider two different scenarios.

In the first scenario, the goal is to compute the planar poses of $n \in \mathbb{N}$ robots $\{\mathbf{p}_1^G, \ldots, \mathbf{p}_n^G\}$ expressed in the global frame G, where $\mathbf{p}_i^G = \left[x_i^G, y_i^G, \theta_i^G\right] \in SE(3)$ for $i \in \{1, \ldots, n\}$, given $m \in \mathbb{N}$ measurements of relative poses between robots. The robots measure the planar pose (position and orientation) of nearby robots expressed on their own reference frame. We let $\mathbf{p}_j^i \in SE(3)$ be the pose of a robot j relative to robot i. This information is represented by a directed graph $\mathscr{G} = (\mathscr{V}, \mathscr{E})$, where the nodes $\mathscr{V} = \{1, \ldots, n\}$ are the robots, and $\mathscr{E}$ contains the m relative measurements, $|\mathscr{E}| = m$. There is an edge $e = (i, j) \in \mathscr{E}$ from i to j if robot i has a relative measurement of the state of robot j. We assume that the measurement graph $\mathscr{G}$ is directed and weakly connected, and that a robot i can exchange data with both, its in and out neighbors $\mathscr{N}_i$ so that the associated communication graph is undirected,

$$\mathcal{N}_i = \{j \mid (i, j) \in \mathcal{E} \text{ or } (j, i) \in \mathcal{E}\}.$$

We let $\mathscr{A} \in \{0, 1, -1\}^{n \times m}$ be the *negative* incidence matrix of the measurement graph,

$$\mathscr{A}_{i,e} = \begin{cases} -1 & \text{if } e = (i, j) \\ 1 & \text{if } e = (j, i), \text{ for } i \in \{1, \ldots, n\}, e \in \{1, \ldots, m\}, \\ 0 & \text{otherwise} \end{cases} \tag{3.1}$$

and we let $\mathscr{W}_{i,j}$ be the Metropolis weights defined in Eq. (A.3) in Appendix A associated to $\mathscr{G}$. The localization problem consists of estimating the states of the n robots from the relative measurements. Any solution can be determined only up to a rotation and a translation, i.e., several equivalent solutions can be obtained depending on the reference frame selected.

As discussed in [4], one of the robots $a \in \mathcal{V}$, e.g., the first one $a = 1$, can be established as an anchor with state $\mathbf{p}_a^a = \mathbf{0}_{3 \times 1}$, and the poses of the non-anchor robots can be expressed relative to the anchor. We call such approaches anchor-based and add the superscript a to their associated variables. We let $\mathcal{V}^a = \mathcal{V} \setminus \{a\}$ be the set of non-anchor nodes and matrix $\mathscr{A}^a \in \{0, 1, -1\}^{n-1 \times m}$ be the result of deleting the row associated to node a from $\mathscr{A}$ in Eq. (3.1). This is the case considered in our first scenario, where we address the anchor-based planar localization problem for the case that the relative measurements are noisy. Each edge $e = (i, j) \in \mathcal{E}$ in the relative measurements graph $\mathscr{G} = (\mathcal{V}, \mathcal{E})$ has associated noisy measurements of the orientation $\mathbf{z}_e^\theta$ and the position $\mathbf{z}_e^{xy}$ of robot j relative to robot i, with associated covariance matrices $\Sigma_{\mathbf{z}_e^\theta}$ and $\Sigma_{\mathbf{z}_e^{xy}}$. We assume that the measurements are independent since they were acquired individually by the robots. The goal is to estimate the robot poses $\hat{\mathbf{p}}_i^a$ of the non-anchor robots $i \in \mathcal{V}^a$ relative to the anchor a from the noisy relative measurements. We assume that the orientations of the robots satisfy $-\pi/2 < \theta_i < \pi/2$ for all $i \in \mathcal{V}$.

In the second scenario, instead of computing planar robot poses, we consider that each robot $i \in \mathcal{V}$ has a p-dimensional state $\mathbf{x}_i \in \mathbb{R}^p$, and that the measurement $\mathbf{z}_e \in \mathbb{R}^p$ associated to an edge $e = (i, j) \in \mathcal{E}$ relates the states of robots i and j as follows

$$\mathbf{z}_e = \mathbf{x}_j - \mathbf{x}_i + \mathbf{v}_e,$$

where $\mathbf{v}_e \sim N\left(\mathbf{0}_{p \times p}, \Sigma_{\mathbf{z}_e}\right)$ is a Gaussian additive noise. Thus, we solve a position localization problem, although the proposed method can be alternatively applied for estimating speeds, accelerations, or current times. In addition, this method can be used in a pose localization scenario, provided that the robots have previously executed an attitude synchronization [25, 30] or a motion coordination [16] strategy to align their orientations. We estimate the states $\hat{\mathbf{x}}_i^{cen}$ of the robots $i \in \mathcal{V}$ relative to the centroid of the states, and compare the solution with a classical anchor-based one $\hat{\mathbf{x}}_i^a$. In the following sections we explain in detail the two scenarios.

3.3 Planar Localization from Noisy Measurements

The problem addressed in this section consists of computing the planar localization of $n \in \mathbb{N}$ robots $\{\mathbf{p}_1^a, \ldots, \mathbf{p}_n^a\}$, where $\mathbf{p}_i^a = \left[x_i^a, y_i^a, \theta_i^a\right]$ for $i \in \{1, \ldots, n\}$, relative to an anchor robot a, given $m \in \mathbb{N}$ noisy measurements of relative poses between robots. There is a single anchor node $a \in \mathcal{V}$ which is placed at the pose $\mathbf{p}_a^a = \mathbf{0}_{3 \times 1}$. By convention, we let the anchor be the first node, $a = 1$, and denote $\mathcal{V}^a = \mathcal{V} \setminus \{a\}$ the set of non-anchor nodes. Each robot gets noisy measurements of the planar pose (position and orientation) of nearby robots to estimate its localization with respect to an anchor node.

Each edge $e = (i, j) \in \mathcal{E}$ in the relative measurements graph $\mathcal{G} = (\mathcal{V}, \mathcal{E})$ has associated noisy measurements of the orientation $\mathbf{z}_e^\theta$ and the position $\mathbf{z}_e^{xy}$ of robot j relative to robot i, with associated covariance matrices $\Sigma_{\mathbf{z}_e^\theta}$ and $\Sigma_{\mathbf{z}_e^{xy}}$. We let $\mathbf{z}_\theta \in \mathbb{R}^m$, $\mathbf{z}_{xy} \in \mathbb{R}^{2m}$, $\Sigma_{\mathbf{z}_\theta} \in \mathbb{R}^{m \times m}$ and $\Sigma_{\mathbf{z}_{xy}} \in \mathbb{R}^{2m \times 2m}$ contain information of the m measurements,

$$\mathbf{z}_\theta = (\mathbf{z}_1^\theta, \ldots, \mathbf{z}_m^\theta)^T, \qquad \mathbf{z}_{xy} = ((\mathbf{z}_1^{xy})^T, \ldots, (\mathbf{z}_m^{xy})^T)^T,$$
$$\Sigma_{\mathbf{z}_\theta} = \mathrm{Diag}(\Sigma_{\mathbf{z}_1^\theta}, \ldots \Sigma_{\mathbf{z}_m^\theta}), \qquad \Sigma_{\mathbf{z}_{xy}} = \mathrm{blkDiag}(\Sigma_{\mathbf{z}_1^{xy}}, \ldots \Sigma_{\mathbf{z}_m^{xy}}).$$

We assume that the measurements are independent since they were acquired individually by the robots. Thus, the goal is that each robot $i \in \mathcal{V}$ estimates its pose $\hat{\mathbf{p}}_i^a$ relative to this anchor.

This problem is solved by using a three-phases strategy [2]

Phase 1: Compute a suboptimal estimate of the robot orientations $\tilde{\theta}_{\mathcal{V}}^a \in \mathbb{R}^n$ relative to the anchor a for all the robots in $\mathcal{V}$;
Phase 2: Express the position measurements $\mathbf{z}_{xy}$ of the robots in terms of the previously computed orientations;
Phase 3: Compute the estimated poses of the robots $\hat{\mathbf{p}}_{\mathcal{V}}^a = ((\hat{\mathbf{x}}_{\mathcal{V}}^a)^T, (\hat{\theta}_{\mathcal{V}}^a)^T)^T$.

During the rest of the section, we study the method and present a distributed implementation.

3.3.1 Centralized Algorithm

Phase 1

During this first phase, an initial estimate of the robot orientations $\tilde{\theta}_{\mathcal{V}^a} \in \mathbb{R}^{n-1}$ relative to the anchor a is obtained. This estimate is computed based exclusively on the orientation measurements $\mathbf{z}_\theta \in \mathbb{R}^m$ with covariance $\Sigma_{\mathbf{z}_\theta} \in \mathbb{R}^{m \times m}$. When the orientation measurements are considered alone and they belong to $\pm \frac{\pi}{2}$, the estimation problem becomes linear, and the estimated solutions are given by the Weighted Least Squares,

$$\tilde{\theta}^a_{\mathcal{V}a} = \Sigma_{\tilde{\theta}^a_{\mathcal{V}a}} \mathscr{A}^a \Sigma_{\mathbf{z}_\theta}^{-1} \mathbf{z}_\theta, \qquad \Sigma_{\tilde{\theta}^a_{\mathcal{V}a}} = \left(\mathscr{A}^a \Sigma_{\mathbf{z}_\theta}^{-1} (\mathscr{A}^a)^T \right)^{-1}, \qquad (3.2)$$

where $\mathscr{A}^a \in \{0, 1, -1\}^{n-1 \times m}$ is the result of deleting the row associated to the anchor a from the incidence matrix $\mathscr{A}$ of the measurement graph in Eq. (3.1). Recall that the orientation of the anchor is set to zero, $\tilde{\theta}^a_i = 0$ for $i = a$. We let $\tilde{\theta}^a_{\mathcal{V}} \in \mathbb{R}^n$ and $\Sigma_{\tilde{\theta}^a_{\mathcal{V}}} \mathbb{R}^{n \times n}$ contain the orientation of all the robots in $\mathcal{V}$, including the anchor a,

$$\tilde{\theta}^a_{\mathcal{V}} = (0, (\tilde{\theta}^a_{\mathcal{V}a})^T)^T, \qquad \Sigma_{\tilde{\theta}^a_{\mathcal{V}}} = \text{Diag}(0, \Sigma_{\tilde{\theta}^a_{\mathcal{V}a}}). \qquad (3.3)$$

Phase 2

Each relative position measurement $\mathbf{z}^{xy}_e$ associated to the edge $e = (i, j)$, was originally expressed in the local coordinates of robot i. During the second phase, these measurements are transformed into a common orientation using the previously computed $\tilde{\theta}^a_{\mathcal{V}}$.

For each edge $e = (i, j) \in \mathcal{E}$ we let $\tilde{R}_e \in \mathbb{R}^{2 \times 2}$ and $\tilde{S}_e \in \mathbb{R}^{2 \times 2}$ be the following matrices associated to the orientation $\tilde{\theta}_i$ of robot i,

$$\tilde{R}_e = \mathscr{R}(\tilde{\theta}^a_i) = \begin{bmatrix} \cos\tilde{\theta}^a_i & -\sin\tilde{\theta}^a_i \\ \sin\tilde{\theta}^a_i & \cos\tilde{\theta}^a_i \end{bmatrix}, \quad \tilde{S}_e = \mathscr{S}(\tilde{\theta}^a_i) = \begin{bmatrix} -\sin\tilde{\theta}^a_i & \cos\tilde{\theta}^a_i \\ -\cos\tilde{\theta}^a_i & -\sin\tilde{\theta}^a_i \end{bmatrix},$$
$$(3.4)$$

and let the block diagonal matrix $\tilde{R} \in \mathbb{R}^{2m \times 2m}$ compile information from the m edges,

$$\tilde{R} = \mathscr{R}(\tilde{\theta}^a_{\mathcal{V}}) = \text{blkDiag}(\tilde{R}_1, \ldots, \tilde{R}_m). \qquad (3.5)$$

The updated pose measurements in the global coordinates $\mathbf{w} \in \mathbb{R}^{2m+(n-1)}$ and their associated covariance $\Sigma_{\mathbf{w}}$ are

$$\mathbf{w} = \begin{bmatrix} \tilde{\mathbf{z}}_{xy} \\ \tilde{\theta}_{\mathcal{V}a} \end{bmatrix} = \begin{bmatrix} \tilde{R} & \mathbf{0} \\ \mathbf{0} & \mathbf{I}_{n-1} \end{bmatrix} \begin{bmatrix} \mathbf{z}_{xy} \\ \tilde{\theta}_{\mathcal{V}a} \end{bmatrix},$$

$$\Sigma_{\mathbf{w}} = \begin{bmatrix} K & J \\ \mathbf{0} & \mathbf{I}_{n-1} \end{bmatrix} \begin{bmatrix} \Sigma_{\mathbf{z}_{xy}} & \mathbf{0} \\ \mathbf{0} & \Sigma_{\tilde{\theta}_{\mathcal{V}a}} \end{bmatrix} \begin{bmatrix} K^T & \mathbf{0} \\ J^T & \mathbf{I}_{n-1} \end{bmatrix}, \qquad (3.6)$$

where $K \in \mathbb{R}^{2m \times 2m}$ and $J \in \mathbb{R}^{2m \times (n-1)}$ are the Jacobians of the transformation with respect to respectively, $\mathbf{z}_{xy}$ and $\tilde{\theta}_{\mathcal{V}a}$,

$$K = \tilde{R}, \text{ and } \quad J_{e,i} = \tilde{S}_e \, \mathbf{z}^{xy}_e \text{ if } e = (i, j) \text{ for some } j, \text{ and } J_{e,i} = \mathbf{0}_{2 \times 1} \text{ otherwise.}$$
$$(3.7)$$

Phase 3

During the last phase, the positions of the robots $\hat{\mathbf{x}}^a_{\mathcal{V}/a} \in \mathbb{R}^{2(n-1)}$ relative to the anchor node a are computed, and an improved version $\hat{\theta}^a_{\mathcal{V}/a} \in \mathbb{R}^{n-1}$ of the previous orientations $\tilde{\theta}^a_{\mathcal{V}/a}$ is obtained. Let $\hat{\mathbf{p}}^a_{\mathcal{V}/a} \in \mathbb{R}^{3(n-1)}$ contain both, the positions and orientations of the non-anchor robots,

$$\hat{\mathbf{p}}^a_{\mathcal{V}/a} = \begin{bmatrix} \hat{\mathbf{x}}^a_{\mathcal{V}/a} \\ \hat{\theta}^a_{\mathcal{V}/a} \end{bmatrix} = \Sigma_{\hat{\mathbf{p}}^a_{\mathcal{V}/a}} B \Sigma^{-1}_{\mathbf{w}} \mathbf{w}, \qquad \Sigma_{\hat{\mathbf{p}}^a_{\mathcal{V}/a}} = \left(B \Sigma^{-1}_{\mathbf{w}} B^T \right)^{-1}, \qquad (3.8)$$

where $B = \text{blkDiag}\left((\mathscr{A}^a \otimes \mathbf{I}_2), \mathbf{I}_{n-1} \right)$, and $\Sigma_{\mathbf{w}}$ and $\mathbf{w}$ are given by (3.6). The estimated poses $\hat{\mathbf{p}}^a_{\mathcal{V}} \in \mathbb{R}^{3n}$ of all the robots in $\mathcal{V}$, including the anchor a, are given by

$$\hat{\mathbf{p}}^a_{\mathcal{V}} = (\mathbf{0}^T_{3\times1}, (\hat{\mathbf{p}}^a_{\mathcal{V}/a})^T)^T, \qquad \Sigma_{\hat{\mathbf{p}}^a_{\mathcal{V}}} = \text{blkDiag}(\mathbf{0}_{3\times3}, \Sigma_{\hat{\mathbf{p}}^a_{\mathcal{V}/a}}). \qquad (3.9)$$

Algorithm

Considering the three phases together, the estimated positions $\hat{\mathbf{x}}^a_{\mathcal{V}/a}$ and orientations $\hat{\theta}^a_{\mathcal{V}/a}$ of the non-anchor robots are

$$\hat{\mathbf{x}}^a_{\mathcal{V}/a} = L^{-1}(\mathscr{A}^a \otimes \mathbf{I}_2)\Upsilon_{\tilde{\mathbf{z}}_{xy}} \left(\mathbf{I}_{2m} + J \Sigma_{\hat{\theta}^a_{\mathcal{V}/a}} J^T \Upsilon_{\tilde{\mathbf{z}}_{xy}} E \right) \tilde{R} \, \mathbf{z}_{xy},$$

$$\hat{\theta}^a_{\mathcal{V}/a} = (\mathscr{A}^a \Sigma^{-1}_{\mathbf{z}_\theta}(\mathscr{A}^a)^T)^{-1}\mathscr{A}^a \Sigma^{-1}_{\mathbf{z}_\theta}\mathbf{z}_\theta + \Sigma_{\hat{\theta}^a_{\mathcal{V}/a}} J^T \Upsilon_{\tilde{\mathbf{z}}_{xy}} E \, \tilde{R} \, \mathbf{z}_{xy}, \quad \text{where} \qquad (3.10)$$

$$\Upsilon_{\tilde{\mathbf{z}}_{xy}} = (\tilde{R} \Sigma_{\mathbf{z}_{xy}} \tilde{R}^T)^{-1}, \qquad\qquad E = (\mathscr{A}^a \otimes \mathbf{I}_2)^T L^{-1}(\mathscr{A}^a \otimes \mathbf{I}_2)\Upsilon_{\tilde{\mathbf{z}}_{xy}} - \mathbf{I}_{2m},$$

$$\Sigma_{\hat{\theta}^a_{\mathcal{V}/a}} = ((\Sigma_{\tilde{\theta}^a_{\mathcal{V}/a}})^{-1} - J^T \Upsilon_{\tilde{\mathbf{z}}_{xy}} E J)^{-1}, \quad L = (\mathscr{A}^a \otimes \mathbf{I}_2)\Upsilon_{\tilde{\mathbf{z}}_{xy}}(\mathscr{A}^a \otimes \mathbf{I}_2)^T, \qquad (3.11)$$

and $\hat{\mathbf{p}}^a_{\mathcal{V}}$ is obtained from the previous expressions as in Eq. (3.9). A full development of these expressions can be found in Appendix B. This localization algorithm can also been used for solving the Simultaneously Localization and Mapping problem of single-robot systems building graph maps [12, 13].

3.3.2 Distributed Algorithm

From (3.10), it can be seen that the computation of $\hat{\mathbf{x}}^a_{\mathcal{V}}$ and $\hat{\theta}^a_{\mathcal{V}}$ involves matrix inversions and other operations that require the knowledge of the whole system. Although, a priori the proposed localization strategy would require a centralized implementation, in the next sections we show a proposal to carry out the computations in a distributed way.

Phase 1

The initial orientation $\tilde{\theta}_{\mathscr{V}}^a$ in the first phase of the algorithm can be computed in a distributed fashion using the following Jacobi algorithm [4]. Let each robot $i \in \mathscr{V}$ maintain a variable $\tilde{\theta}_i^a(t) \in \mathbb{R}$. The anchor $i = a$ keeps its variable equal to zero for all time steps $t \in \mathbb{N}$,

$$\tilde{\theta}_i^a(0) = 0, \qquad \tilde{\theta}_i^a(t+1) = \tilde{\theta}_i^a(t), \qquad \text{for } i = a. \qquad (3.12)$$

Each non-anchor robot $i \in \mathscr{V}^a$ initializes its variable at $t = 0$ with any value $\tilde{\theta}_i^a(0)$, and updates it at each time step $t \in \mathbb{N}$ by

$$\tilde{\theta}_i^a(t+1) = C_i^{-1}c_i + C_i^{-1} \sum_{e=(i,j)\in\mathscr{E}} (\Sigma_{\mathbf{z}_e^\theta})^{-1}\tilde{\theta}_j^a(t) + C_i^{-1} \sum_{e=(j,i)\in\mathscr{E}} (\Sigma_{\mathbf{z}_e^\theta})^{-1}\tilde{\theta}_j^a(t),$$

$$(3.13)$$

where

$$c_i = -\sum_{e=(i,j)\in\mathscr{E}} (\Sigma_{\mathbf{z}_e^\theta})^{-1}\mathbf{z}_e^\theta + \sum_{e=(j,i)\in\mathscr{E}} (\Sigma_{\mathbf{z}_e^\theta})^{-1}\mathbf{z}_e^\theta,$$

$$C_i = \sum_{e=(i,j)\in\mathscr{E}} (\Sigma_{\mathbf{z}_e^\theta})^{-1} + \sum_{e=(j,i)\in\mathscr{E}} (\Sigma_{\mathbf{z}_e^\theta})^{-1}. \qquad (3.14)$$

The previous expressions are the Jacobi iterations associated to (3.2). Let $\Upsilon_{\tilde{\theta}_{\mathscr{V}a}^a}$ and $\eta_{\tilde{\theta}_{\mathscr{V}a}^a}$ be respectively the information matrix and vector of $\tilde{\theta}_{\mathscr{V}a}^a$,

$$\Upsilon_{\tilde{\theta}_{\mathscr{V}a}^a} = (\Sigma_{\tilde{\theta}_{\mathscr{V}a}^a})^{-1} = \mathscr{A}^a \Sigma_{\mathbf{z}_\theta}^{-1}(\mathscr{A}^a)^T, \qquad \eta_{\tilde{\theta}_{\mathscr{V}a}^a} = \mathscr{A}^a \Sigma_{\mathbf{z}_\theta}^{-1}\mathbf{z}_\theta. \qquad (3.15)$$

Let C contain the elements in the diagonal of $\Upsilon_{\tilde{\theta}_{\mathscr{V}a}^a}$,

$$C = \text{Diag}([\Upsilon_{\tilde{\theta}_{\mathscr{V}a}^a}]_{2,2}, \ldots, [\Upsilon_{\tilde{\theta}_{\mathscr{V}a}^a}]_{n,n}),$$

and D be $D = C - \Upsilon_{\tilde{\theta}_{\mathscr{V}a}^a}$. The first equation in (3.2) can be rewritten as

$$\Upsilon_{\tilde{\theta}_{\mathscr{V}a}^a}\tilde{\theta}_{\mathscr{V}a}^a = \eta_{\tilde{\theta}_{\mathscr{V}a}^a}, \qquad \tilde{\theta}_{\mathscr{V}a}^a = C^{-1}D\tilde{\theta}_{\mathscr{V}a}^a + C^{-1}\eta_{\tilde{\theta}_{\mathscr{V}a}^a}. \qquad (3.16)$$

From here, we can write

$$\tilde{\theta}_{\mathscr{V}a}^a(t+1) = C^{-1}D\tilde{\theta}_{\mathscr{V}a}^a(t) + C^{-1}\eta_{\tilde{\theta}_{\mathscr{V}a}^a}, \qquad (3.17)$$

initialized at $t = 0$ with $\tilde{\theta}_{\mathscr{V}a}^a(0)$. By operating with $\mathscr{A}^a \Sigma_{\mathbf{z}_\theta}^{-1}\mathbf{z}_\theta$ and $\mathscr{A}^a \Sigma_{\mathbf{z}_\theta}^{-1}(\mathscr{A}^a)^T$, it can be seen that (3.13) is the ith row of (3.17). The system (3.17) converges to

$\tilde{\theta}^a_{\mathcal{V}a}$ in Eq. (3.2), and equivalently each $\tilde{\theta}^a_i(t)$ in (3.13) converges to $\tilde{\theta}^a_i$ for $i \in \mathcal{V}^a$, if the spectral radius of $C^{-1}D$ is less than 1,

$$\rho(C^{-1}D) < 1, \tag{3.18}$$

and the anchor variable, $\tilde{\theta}^a_i(t)$ with $i = a$, remains equal to 0 for all the iterations t. The value $\rho(C^{-1}D)$ gives the convergence speed of the system, converging faster for $\rho(C^{-1}D)$ closer to 0. Recalling that $\Sigma_{\mathbf{z}_\theta}$ is a diagonal matrix, then each variable $\tilde{\theta}^a_i(t)$ asymptotically converges to the ith entry $\tilde{\theta}^a_i$ of the vector $\tilde{\theta}^a_{\mathcal{V}a}$ in (3.2) [4] that would be computed by a centralized system.

Observe that the computations are fully distributed and they exclusively rely on local information. The constants C_i and c_i are computed by each robot $i \in \mathcal{V}^a$ using exclusively the measurements $\mathbf{z}^\theta_e$ and covariances $\Sigma_{\mathbf{z}^\theta_e}$ of its incoming $e = (j, i)$ or outgoing edges $e = (i, j)$. Also the variables $\tilde{\theta}^a_j(t)$ used to update its own $\tilde{\theta}^a_i(t + 1)$ belong to neighboring robots $j \in \mathcal{N}_i$.

Phase 2

Let us assume that the robots have executed $t_{\max}$ iterations of the previous algorithm, and let $\bar{\theta}^a_i$ be their orientation at iteration $t_{\max}$, $\bar{\theta}^a_i = \tilde{\theta}^a_i(t_{\max})$. Then, the second phase of the algorithm is executed to transform the locally expressed measurements $\mathbf{z}_{xy}$ into the measurements expressed in the reference frame of the anchor node $\tilde{\mathbf{z}}_{xy}$. As previously stated, the estimated orientations $\bar{\theta}^a_i$ do not change during this phase (3.6). Let $\bar{R} = \mathcal{R}(\bar{\theta}^a_{\mathcal{V}a})$ be defined by using the orientations $\bar{\theta}^a_i$ instead of $\tilde{\theta}^a_i$ in (3.5). Since the matrix $\bar{R}$ is block diagonal, each robot $i \in \mathcal{V}$ can locally transform its own local measurements,

$$\bar{\mathbf{z}}^{xy}_e = \bar{R}_e \mathbf{z}^{xy}_e, \text{ for all } e = (i, j) \in \mathcal{E}. \tag{3.19}$$

Since the robots use $\bar{\theta}$ instead of $\tilde{\theta}$, also the updated measurements obtained during the second phase are $\bar{\mathbf{z}}_{xy}$ instead of $\tilde{\mathbf{z}}_{xy}$. This second phase is local and it is executed in a single iteration.

Phase 3

In order to obtain the final estimate $\hat{\mathbf{p}}^a_{\mathcal{V}a}$, the third step of the algorithm (3.8) apparently requires the knowledge of the covariance matrix $\Sigma_\mathbf{w}$, which at the same time, requires the knowledge of $\Sigma_{\bar{\theta}^a_{\mathcal{V}a}}$. However, a distributed computation of these matrices cannot be carried out in an efficient way. Here we present a distributed algorithm for computing $\hat{\mathbf{p}}^a_{\mathcal{V}a}$.

Let each robot $i \in \mathcal{V}$ maintain a variable $\hat{\mathbf{p}}^a_i(t) \in \mathbb{R}^3$, composed of its estimated position $\hat{\mathbf{x}}^a_i(t) \in \mathbb{R}^2$ and orientation $\hat{\theta}^a_i(t) \in \mathbb{R}$, and let $\hat{\mathbf{p}}^a_{\mathcal{V}}(t)$ be the result of putting together the $\mathbf{p}^a_i(t)$ variables for all $i \in \mathcal{V}$. The anchor robot keeps its variable equal to zero for all the iterations,

$$\hat{\mathbf{p}}_i^a(0) = \mathbf{0}_{3\times 1}, \qquad \hat{\mathbf{p}}_i^a(t+1) = \hat{\mathbf{p}}_i^a(t), \qquad \text{for } i = a. \tag{3.20}$$

Each non-anchor robot $i \in \mathcal{V}^a$ initializes its variable at $t = 0$ with any value $\hat{\mathbf{p}}_i^a(0)$ and updates $\hat{\mathbf{p}}_i^a(t)$ at each time step $t \in \mathbb{N}$ by

$$\hat{\mathbf{p}}_i^a(t+1) = \begin{bmatrix} \hat{\mathbf{x}}_i^a(t+1) \\ \hat{\theta}_i^a(t+1) \end{bmatrix} = M_i^{-1}\left(\mathbf{f}_i(\hat{\mathbf{p}}_{\mathcal{V}}^a(t)) + \mathbf{m}_i\right), \tag{3.21}$$

where

$$M_i = \begin{bmatrix} M_1 & M_2 \\ M_3 & M_4 \end{bmatrix}, \ \mathbf{f}_i(\mathbf{p}_{\mathcal{V}}^a(t)) = \begin{bmatrix} f_1 \\ f_2 \end{bmatrix}, \ \mathbf{m}_i = \begin{bmatrix} m_1 \\ m_2 \end{bmatrix}. \tag{3.22}$$

Let $\Upsilon_{\tilde{\mathbf{z}}_e^{xy}}$ be the block within the matrix $\Upsilon_{\tilde{\mathbf{z}}_{xy}}$ in (3.11) associated to an edge $e = (i,j) \in \mathcal{E}$,

$$\Upsilon_{\tilde{\mathbf{z}}_e^{xy}} = \tilde{R}_e(\Sigma_{\mathbf{z}_e^{xy}})^{-1}(\tilde{R}_e)^T. \tag{3.23}$$

The elements within M_i are

$$M_1 = \sum_{e=(i,j)\in\mathcal{E}} \Upsilon_{\tilde{\mathbf{z}}_e^{xy}} + \sum_{e=(j,i)\in\mathcal{E}} \Upsilon_{\tilde{\mathbf{z}}_e^{xy}},$$

$$M_2 = \sum_{e=(i,j)\in\mathcal{E}} \Upsilon_{\tilde{\mathbf{z}}_e^{xy}} \tilde{S}_e \mathbf{z}_e^{xy},$$

$$M_3 = \sum_{e=(i,j)\in\mathcal{E}} (\mathbf{z}_e^{xy})^T (\tilde{S}_e)^T \Upsilon_{\tilde{\mathbf{z}}_e^{xy}},$$

$$M_4 = \sum_{e=(i,j)\in\mathcal{E}} (\mathbf{z}_e^{xy})^T (\tilde{S}_e)^T \Upsilon_{\tilde{\mathbf{z}}_e^{xy}} \tilde{S}_e \mathbf{z}_e^{xy} + \sum_{e=(i,j)\in\mathcal{E}} (\Sigma_{\mathbf{z}_e^{xy}})^{-1} + \sum_{e=(j,i)\in\mathcal{E}} (\Sigma_{\mathbf{z}_e^{xy}})^{-1}. \tag{3.24}$$

The elements within $\mathbf{f}_i(\hat{\mathbf{p}}_{\mathcal{V}}^a(t))$, which is the term depending on the previous estimates $\hat{\mathbf{p}}_{\mathcal{V}}^a(t) = (\hat{\mathbf{x}}_{\mathcal{V}}^a(t)^T, \hat{\theta}(t)_{\mathcal{V}}^a)^T$, are

$$f_1 = \sum_{e=(i,j)\in\mathcal{E}} \Upsilon_{\tilde{\mathbf{z}}_e^{xy}} \hat{\mathbf{x}}_j^a(t) + \sum_{e=(j,i)\in\mathcal{E}} \Upsilon_{\tilde{\mathbf{z}}_e^{xy}} \hat{\mathbf{x}}_j^a(t) + \sum_{e=(j,i)\in\mathcal{E}} \Upsilon_{\tilde{\mathbf{z}}_e^{xy}} \tilde{S}_e \mathbf{z}_e^{xy} \hat{\theta}_j^a(t),$$

$$f_2 = \sum_{e=(i,j)\in\mathcal{E}} (\mathbf{z}_e^{xy})^T (\tilde{S}_e)^T \Upsilon_{\tilde{\mathbf{z}}_e^{xy}} \hat{\mathbf{x}}_j^a(t) - \sum_{e=(i,j)\in\mathcal{E}} (\Sigma_{\mathbf{z}_e^{xy}})^{-1}\hat{\theta}_j^a(t) - \sum_{e=(j,i)\in\mathcal{E}} (\Sigma_{\mathbf{z}_e^{xy}})^{-1}\hat{\theta}_j^a(t). \tag{3.25}$$

Finally, the terms within $\mathbf{m}_i$ are

$$
m_1 = - \sum_{e=(i,j)\in\mathscr{E}} \Upsilon_{\tilde{\mathbf{z}}_e^{xy}} \tilde{\mathbf{z}}_e^{xy} + \sum_{e=(j,i)\in\mathscr{E}} \Upsilon_{\tilde{\mathbf{z}}_e^{xy}} \tilde{\mathbf{z}}_e^{xy}
$$

$$
+ \sum_{e=(i,j)\in\mathscr{E}} \Upsilon_{\tilde{\mathbf{z}}_e^{xy}} \tilde{S}_e \mathbf{z}_e^{xy} \tilde{\theta}_i^a - \sum_{e=(j,i)\in\mathscr{E}} \Upsilon_{\tilde{\mathbf{z}}_e^{xy}} \tilde{S}_e \mathbf{z}_e^{xy} \tilde{\theta}_j^a,
$$

$$
m_2 = - \sum_{e=(i,j)\in\mathscr{E}} (\mathbf{z}_e^{xy})^T (\tilde{S}_e)^T \Upsilon_{\tilde{\mathbf{z}}_e^{xy}} \tilde{\mathbf{z}}_e^{xy} + \sum_{e=(i,j)\in\mathscr{E}} (\mathbf{z}_e^{xy})^T (\tilde{S}_e)^T \Upsilon_{\tilde{\mathbf{z}}_e^{xy}} \tilde{S}_e \mathbf{z}_e^{xy} \tilde{\theta}_i^a
$$

$$
- \sum_{e=(i,j)\in\mathscr{E}} (\Sigma_{\mathbf{z}_e^\theta})^{-1} \tilde{\theta}_j^a - \sum_{e=(j,i)\in\mathscr{E}} (\Sigma_{\mathbf{z}_e^\theta})^{-1} \tilde{\theta}_j^a
$$

$$
+ \sum_{e=(i,j)\in\mathscr{E}} (\Sigma_{\mathbf{z}_e^\theta})^{-1} \tilde{\theta}_i^a + \sum_{e=(j,i)\in\mathscr{E}} (\Sigma_{\mathbf{z}_e^\theta})^{-1} \tilde{\theta}_i^a. \tag{3.26}
$$

Theorem 5 *The estimates $\hat{\mathbf{p}}_i^a(t)$ computed by each robot $i \in \mathscr{V}$ by the distributed algorithm (3.20)–(3.21) converge to $\hat{\mathbf{p}}_i^a = [(\hat{\mathbf{x}}_i^a)^T \hat{\theta}_i^a]^T$ for connected measurement graphs $\mathscr{G}$ with ring or string structure.*

Proof For the anchor $i = a$, it is true since $\hat{\mathbf{p}}_i^a(t) = \mathbf{0}$ for all the time steps. Now we focus on the non-anchor nodes in $\mathscr{V}^a$. First of all, we show that $\hat{\mathbf{p}}_i^a$ is an equilibrium point of the algorithm (3.21) for all $i \in \mathscr{V}^a$. Let $\Upsilon_{\hat{\mathbf{p}}_{\mathscr{V}/a}^a}$ be the information matrix associated to $\hat{\mathbf{p}}_{\mathscr{V}/a}^a$, i.e., $\Upsilon_{\hat{\mathbf{p}}_{\mathscr{V}/a}^a} = (\Sigma_{\hat{\mathbf{p}}_{\mathscr{V}/a}^a})^{-1}$,

$$
\Upsilon_{\hat{\mathbf{p}}_{\mathscr{V}/a}^a} = \begin{bmatrix} L & -\mathscr{A}^a \Upsilon_{\tilde{\mathbf{z}}_{xy}} J \\ -J^T \Upsilon_{\tilde{\mathbf{z}}_{xy}} (\mathscr{A}^a \otimes \mathbf{I}_2)^T & \mathscr{A}^a \Sigma_{\mathbf{z}_\theta}^{-1} (\mathscr{A}^a)^T + J^T \Upsilon_{\tilde{\mathbf{z}}_{xy}} J \end{bmatrix}, \tag{3.27}
$$

where L and $\Upsilon_{\tilde{\mathbf{z}}_{xy}}$ are given by (3.11). Analyzing the term $B\Sigma_{\mathbf{w}}^{-1}$ in (3.8), it can be seen that

$$
B\Sigma_{\mathbf{w}}^{-1} = \begin{bmatrix} (\mathscr{A}^a \otimes \mathbf{I}_2) \Upsilon_{\tilde{\mathbf{z}}_{xy}} & -(\mathscr{A}^a \otimes \mathbf{I}_2) \Upsilon_{\tilde{\mathbf{z}}_{xy}} J \\ -J^T \Upsilon_{\tilde{\mathbf{z}}_{xy}} & \mathscr{A}^a \Upsilon_{\mathbf{z}_\theta}^{-1} (\mathscr{A}^a \otimes \mathbf{I}_2)^T + J^T \Upsilon_{\tilde{\mathbf{z}}_{xy}} J \end{bmatrix}. \tag{3.28}
$$

If we express the third phase in the following way

$$
\Upsilon_{\hat{\mathbf{p}}_{\mathscr{V}/a}^a} \hat{\mathbf{p}}_{\mathscr{V}/a}^a = B \Sigma_{\mathbf{w}}^{-1} \mathbf{w}, \tag{3.29}
$$

and then we consider the rows associated to robot i, we get

$$
\hat{\mathbf{p}}_i^a = \begin{bmatrix} \hat{\mathbf{x}}_i^a \\ \hat{\theta}_i^a \end{bmatrix} = M_i^{-1} \left(\mathbf{f}_i(\hat{\mathbf{p}}_{\mathscr{V}}^a) + \mathbf{m}_i \right), \tag{3.30}
$$

with M_i, $\mathbf{f}_i(\mathbf{p}_{\mathscr{V}}^a(t))$ and $\mathbf{m}_i$ as in (3.22)–(3.26).

Now we prove that the system is convergent. Let $M = \text{blkDiag}(M_2, \ldots, M_n)$ and $\hat{\mathbf{q}}^a_{\mathscr{V}^a}$ be a permutation of $\hat{\mathbf{p}}^a_{\mathscr{V}^a}$ so that the estimates of each robot appear together, $\hat{\mathbf{q}}^a_{\mathscr{V}^a} = \left[(\hat{\mathbf{x}}^a_2)^T \hat{\theta}^a_2, \ldots, (\hat{\mathbf{x}}^a_n)^T \hat{\theta}^a_n \right]^T$. Equivalently, the permuted version of the information matrix $\Upsilon_{\hat{\mathbf{p}}^a_{\mathscr{V}^a}}$ is $\Upsilon_{\hat{\mathbf{q}}^a_{\mathscr{V}^a}}$. The estimates $\hat{\mathbf{p}}^a_i(t)$ computed by each robot $i \in \mathscr{V}^a$ with the distributed algorithm (3.21) converge to $\hat{\mathbf{p}}^a_i = [(\hat{\mathbf{x}}^a_i)^T \hat{\theta}^a_i]^T$ if $\rho(M^{-1}(M - \Upsilon_{\hat{\mathbf{q}}^a_{\mathscr{V}^a}})) < 1$, or equivalently if

$$\rho(\mathbf{I} - M^{-1}\Upsilon_{\hat{\mathbf{q}}^a_{\mathscr{V}^a}}) < 1. \tag{3.31}$$

Since $\lambda(\mathbf{I} - M^{-1}\Upsilon_{\hat{\mathbf{q}}^a_{\mathscr{V}^a}}) = 1 - \lambda(M^{-1}\Upsilon_{\hat{\mathbf{q}}^a_{\mathscr{V}^a}})$, then (3.21) converges if $0 < \lambda(M^{-1}\Upsilon_{\hat{\mathbf{q}}^a_{\mathscr{V}^a}}) < 2$. The first part $0 < \lambda(M^{-1}\Upsilon_{\hat{\mathbf{q}}^a_{\mathscr{V}^a}})$ can be easily checked taking into account that both M^{-1} and $\Upsilon_{\hat{\mathbf{q}}^a_{\mathscr{V}^a}}$ are nonsingular, symmetric, positive definite, and that $\lambda(M^{-1}\Upsilon_{\hat{\mathbf{q}}^a_{\mathscr{V}^a}}) \geq \frac{\lambda_{\min}(M^{-1})}{\lambda_{\max}(\Upsilon_{\hat{\mathbf{q}}^a_{\mathscr{V}^a}})}$ [23, Lemma 1]. Since $0 < \frac{\lambda_{\min}(M^{-1})}{\lambda_{\max}(\Upsilon_{\hat{\mathbf{q}}^a_{\mathscr{V}^a}})}$, then $0 < \lambda(M^{-1}\Upsilon_{\hat{\mathbf{q}}^a_{\mathscr{V}^a}})$.

In order to prove the second part, $\lambda(M^{-1}\Upsilon_{\hat{\mathbf{q}}^a_{\mathscr{V}^a}}) < 2$, let us first focus on the structure of the information matrix $\Upsilon_{\hat{\mathbf{q}}^a_{\mathscr{V}^a}}$. This matrix has zeros for the elements associated to non neighboring robots, and thus it is compatible with $\text{adj}(\mathscr{G}) \otimes \mathbf{I}_3$, where $\text{adj}(\mathscr{G})$ is the adjacency matrix of the graph, and $\mathbf{I}_3$ is the 3×3 identity matrix. For ring or string graphs, the adjacency matrix can be reordered grouping the elements around the main diagonal resulting in a matrix that has semi bandwidth $s = 1$, i.e.,

$$\text{adj}(\mathscr{G})_{ij} = 0 \text{ for } |i - j| > s.$$

As a consequence, the information matrix $\Upsilon_{\hat{\mathbf{q}}^a_{\mathscr{V}^a}}$ has block semi bandwidth $s' = 1$, and as stated by [23, Theorem 1],

$$\lambda_{\max}(M^{-1}\Upsilon_{\hat{\mathbf{q}}^a_{\mathscr{V}^a}}) < 2^{s'} = 2.$$

$\square$

Due to the structure of the information matrices, the third phase of the algorithm can be expressed in terms of local information (3.21)–(3.26) and interactions with neighbors, and thus it can be implemented in a distributed fashion. It is observed that the robots actually use $\bar{\theta}^a_{\mathscr{V}}$ instead of $\tilde{\theta}^a_{\mathscr{V}}$ and as a result, the solution obtained is slightly different from the one in the centralized case. We experimentally analyze the effects of these differences later in this chapter.

3.4 Centroid-Based Position Estimation from Noisy Measurements

This section discusses a higher dimensional scenario. We addresses the problem of estimation of position from noisy measurements of the relative positions of neighbors. The method simultaneously estimates the centroid of the network. Each robot in the network obtains its three dimensional position relative to the estimated centroid. The usual approaches to multi-robot localization assume instead that one anchor robot exists in the network, and the other robots positions are estimated with respect to the anchor. We show that the studied centroid-based algorithm converges to the optimal solution, and that such a centroid-based representation produces results that are more accurate than anchor-based ones, irrespective of the selected anchor [3]. In previous sections we denoted $\mathbf{p}_i$ the pose of a robot i. Since in this section we exclusively consider robot positions, for clarity we use a different symbol $\mathbf{x}_i$ for the robots variables.

Consider that each robot $i \in \{1, \ldots, n\}$ has a p−dimensional state $\mathbf{x}_i \in \mathbb{R}^p$ and it observes the states of a subset of the robots relative to its own state, $\mathbf{x}_j - \mathbf{x}_i$. These states can be positions in cartesian coordinates or, in other situations, orientations, speeds, accelerations, or current times. Each edge $e = (i, j) \in \mathcal{E}$ in the relative measurements graph $\mathcal{G} = (\mathcal{V}, \mathcal{E})$ represents that robot i has a noisy relative measurement $\mathbf{z}_e \in \mathbb{R}^p$ of the state of robot j,

$$\mathbf{z}_e = \mathbf{x}_j - \mathbf{x}_i + \mathbf{v}_e, \tag{3.32}$$

where $\mathbf{v}_e \sim N\left(\mathbf{0}_{p \times p}, \Sigma_{\mathbf{z}_e}\right)$ is a Gaussian additive noise. We let $\mathbf{z} \in \mathbb{R}^{mp}$ and $\Sigma_{\mathbf{z}} \in \mathbb{R}^{mp \times mp}$ contain the information of the m measurements,

$$\mathbf{z} = (\mathbf{z}_1^T, \ldots, \mathbf{z}_m^T)^T, \qquad \Sigma_{\mathbf{z}} = \text{blkDiag}(\Sigma_{\mathbf{z}_1}, \ldots, \Sigma_{\mathbf{z}_m}). \tag{3.33}$$

We assume that the measurement graph $\mathcal{G}$ is directed and weakly connected, and that an robot i can exchange data with both its in and out neighbors $\mathcal{N}_i$ so that the associated communication graph is undirected. The estimation from relative measurements problem consists of estimating the states of the n robots from the relative measurements $\mathbf{z}$. Any solution can be determined only up to an additive constant. Conventionally [4] one of the robots $a \in \mathcal{V}$, e.g., the first one $a = 1$, is established as an anchor with state $\hat{\mathbf{x}}_a^a = \mathbf{0}_p$. We call such approaches anchor-based and add the superscript a to their associated variables. The Best Linear Unbiased Estimator of the states $\hat{\mathbf{x}}_{\mathcal{V}^a}^a \in \mathbb{R}^{(n-1)p}$, $\hat{\mathbf{x}}_{\mathcal{V}^a}^a = ((\hat{\mathbf{x}}_2^a)^T, \ldots, (\hat{\mathbf{x}}_n^a)^T)^T$, of the non-anchor robots $\mathcal{V}^a = \mathcal{V} \setminus \{a\}$ relative to a are obtained as follows [4],

$$\hat{\mathbf{x}}_{\mathcal{V}^a}^a = \Sigma_{\hat{\mathbf{x}}_{\mathcal{V}^a}^a} \left(\mathscr{A}^a \otimes \mathbf{I}_p\right) \Sigma_{\mathbf{z}}^{-1} \mathbf{z}, \quad \Sigma_{\hat{\mathbf{x}}_{\mathcal{V}^a}^a} = \left((\mathscr{A}^a \otimes \mathbf{I}_p) \Sigma_{\mathbf{z}}^{-1} (\mathscr{A}^a \otimes \mathbf{I}_p)^T\right)^{-1}, \tag{3.34}$$

where $\mathscr{A}^a \in \mathbb{R}^{(n-1) \times m}$ is the incidence matrix of $\mathscr{G}$ as in Eq. (3.1), but without the row associated to the anchor a. From now on, both $\hat{\mathbf{x}}_{\mathscr{V}}^a = (\mathbf{0}_p^T, (\hat{\mathbf{x}}_{\mathscr{V}a}^a)^T)^T$ and $\Sigma_{\hat{\mathbf{x}}_{\mathscr{V}}^a} = \text{blkDiag}\left(\mathbf{0}_{p \times p}, \Sigma_{\hat{\mathbf{x}}_{\mathscr{V}a}^a}\right)$, include the estimated state of the anchor a as well.

3.4.1 Position Estimation Relative to an Anchor

We present first distributed strategies where each robot i iteratively estimates its own position relative to an anchor through local interactions with its neighbors $\mathcal{N}_i$. Among the different existing methods for estimating the states $\hat{\mathbf{x}}_{\mathscr{V}}^a$ relative to an anchor, we present the Jacobi algorithm [4], although other distributed methods such as the Jacobi Overrelaxation [8], or the Overlapping Subgraph Estimator [5] could alternatively be applied. The approach in [28], based on the cycle structure of the graph, could be used as well, although it requires multi-hop communication.

Considering Eq. (3.34), it can be seen that computing $\hat{\mathbf{x}}_{\mathscr{V}a}^a$ is equivalent to finding a solution to the system $\Upsilon \hat{\mathbf{x}}_{\mathscr{V}a}^a = \eta$, being η and Υ the information vector and matrix associated to $\hat{\mathbf{x}}_{\mathscr{V}a}^a$ and $\Sigma_{\hat{\mathbf{x}}_{\mathscr{V}a}^a}$,

$$\eta = \left(\mathscr{A}^a \otimes \mathbf{I}_p\right) \Sigma_{\mathbf{z}}^{-1} \mathbf{z}, \qquad \Upsilon = \left(\mathscr{A}^a \otimes \mathbf{I}_p\right) \Sigma_{\mathbf{z}}^{-1} \left(\mathscr{A}^a \otimes \mathbf{I}_p\right)^T. \qquad (3.35)$$

This can be iteratively solved with the Jacobi method [8], where the variable $\hat{\mathbf{x}}_{\mathscr{V}a}^a(t) \in \mathbb{R}^{(n-1)p}$ is initialized with an arbitrary value $\hat{\mathbf{x}}_{\mathscr{V}a}^a(0)$ and it is updated at each step t with the following rule,

$$\hat{\mathbf{x}}_{\mathscr{V}a}^a(t+1) = D^{-1} N \hat{\mathbf{x}}_{\mathscr{V}a}^a(t) + D^{-1} \eta, \qquad (3.36)$$

being D, N the following decomposition of $\Upsilon = [\Upsilon_{ij}]$:

$$D = \text{blkDiag}(\Upsilon_{22}, \dots, \Upsilon_{nn}), \qquad N = D - \Upsilon. \qquad (3.37)$$

The previous variable $\hat{\mathbf{x}}_{\mathscr{V}a}^a(t)$ converges to $\hat{\mathbf{x}}_{\mathscr{V}a}^a$ if the Jacobi matrix $J = D^{-1} N$ has spectral radius less than or equal to one, $\rho(J) = \rho(D^{-1}N) < 1$. The interest of the Jacobi method is that it can be executed in a distributed fashion when the information matrix Υ is compatible with the graph (if $j \notin \mathcal{N}_i$ then $\Upsilon_{ij} = \Upsilon_{ji} = \mathbf{0}_{p \times p}$), and when in addition the rows of Υ and of η associated to each robot $i \in \mathscr{V}^a$ only depend on data which is local to robot i. Next, the general anchor-based estimation algorithm [4] based on the Jacobi method is presented. It allows each robot $i \in \mathscr{V}$ to iteratively estimate its own $\hat{\mathbf{x}}_i^a$ within $\hat{\mathbf{x}}_{\mathscr{V}a}^a = ((\hat{\mathbf{x}}_2^a)^T, \dots, (\hat{\mathbf{x}}_n^a)^T)^T$ in a distributed fashion.

Algorithm 3 Let each robot $i \in \mathscr{V}$ have a variable $\hat{\mathbf{x}}_i^a(t) \in \mathbb{R}^p$ initialized at $t = 0$ with $\hat{\mathbf{x}}_i^a(0) = \mathbf{0}_p$. At each time step t, each robot $i \in \mathscr{V}$ updates $\hat{\mathbf{x}}_i^a(t)$ with

$$\hat{\mathbf{x}}_i^a(t+1) = \sum_{j\in\mathcal{N}_i} M_i \mathcal{B}_{ij} \hat{\mathbf{x}}_j^a(t) + \sum_{e=(j,i)\in\mathcal{E}} M_i \Sigma_{\mathbf{z}_e}^{-1} \mathbf{z}_e - \sum_{e=(i,j)\in\mathcal{E}} M_i \Sigma_{\mathbf{z}_e}^{-1} \mathbf{z}_e,$$

$$(3.38)$$

where M_i and $\mathcal{B}_{ij}$ are $p \times p$ matrices with $M_i = \mathbf{0}$ for $i = a$, $M_i = (\sum_{j\in\mathcal{N}_i} \mathcal{B}_{ij})^{-1}$ for $i \neq a$, and

$$\mathcal{B}_{ij} = \begin{cases} \Sigma_{\mathbf{z}_e}^{-1} + \Sigma_{\mathbf{z}_{e'}}^{-1} & \text{if } e = (i,j), e' = (j,i) \in \mathcal{E} \\ \Sigma_{\mathbf{z}_e}^{-1} & \text{if } e = (i,j) \in \mathcal{E}, (j,i) \notin \mathcal{E} \\ \Sigma_{\mathbf{z}_e}^{-1} & \text{if } e = (j,i) \in \mathcal{E}, (i,j) \notin \mathcal{E} \end{cases} \qquad (3.39)$$

The convergence of this estimation algorithm has been proved [4, Theorem 1] for connected measurement graphs with independent relative measurements, under the assumption that either

(i) The covariance matrices of the measurements are exactly diagonal; or
(ii) All measurements have exactly the same covariance matrix.

However, we would like the algorithm presented here to be applicable to a wider case of relative noises, in particular to independent noises, with not necessarily diagonal or equal covariance matrices. Next we use results on block matrices [17], see Appendix B, to prove the convergence of the Jacobi algorithm for this more general case.

Theorem 6 *Let the measurement graph $\mathcal{G}$ be weakly connected, $\Sigma_{\mathbf{z}_1}, \ldots, \Sigma_{\mathbf{z}_m}$ be the covariance matrices, not necessarily equal or diagonal, associated to m independent p–dimensional measurements, and $\Sigma_{\mathbf{z}}$ be their associated block-diagonal covariance matrix as in Eq. (3.33). Then, the spectral radius of $D^{-1}N$, with D and N computed as in Eqs. (3.35)–(3.37), is less than 1,*

$$\rho(D^{-1}N) < 1. \qquad (3.40)$$

Proof In order to prove (3.40) we use the definitions and results in Appendix B. We first analyze the contents of Υ and show that Υ is of class Z_{n-1}^p according to Definition 6 in Appendix B. Then, we use Lemma 4 and Theorem 9 to show that Υ is of class M_{n-1}^p as in Definition 6. Finally, we show that $\Upsilon + \Upsilon^T \in M_{n-1}^p$ and use Theorem 10 to prove (3.40). Note that the subscript $n-1$ used in this proof instead of n comes from the fact that $\Upsilon = [\Upsilon_{ij}]$, with $i, j \in \mathcal{V}^a$ and $|\mathcal{V}^a| = n-1$.

We first analyze the contents of the information matrix Υ given by Eq. (3.35). Each block Υ_{ij} of the information matrix Υ is given by

$$\Upsilon_{ij} = \begin{cases} -\mathcal{B}_{ij} & \text{if } j \in \mathcal{N}_i, j \neq i \\ \mathbf{0} & \text{if } j \notin \mathcal{N}_i, j \neq i \end{cases}, \qquad \text{and } \Upsilon_{ii} = \sum_{j\in\mathcal{N}_i} \mathcal{B}_{ij}, \qquad (3.41)$$

for $i, j \in \mathcal{V}^a$, where $\mathcal{B}_{ij}$ is given by Eq. (3.39). Note that $\mathcal{B}_{ij}$ is symmetric and that $\mathcal{B}_{ij} \succ \mathbf{0}$[1] and thus $-\mathcal{B}_{ij} \prec \mathbf{0}$ and symmetric. Therefore, matrix Υ is of class Z_{n-1}^p according to Definition 6.

Now we focus on Lemma 4. We are interested in showing that, given any subset of robots $\mathcal{J} \subset \mathcal{V}^a$, there exists $i \in \mathcal{J}$ such that $\sum_{j \in \mathcal{J}} \Upsilon_{ij} \succ \mathbf{0}$. First we analyze the case $\mathcal{J} = \mathcal{V}^a$. Observe that Υ does not have any rows or columns associated to the anchor robot a, i.e., $\Upsilon = [\Upsilon_{ij}]$ with $i, j \in \mathcal{V}^a$. On the other hand, for each robot i that has the anchor a as a neighbor, $a \in \mathcal{N}_i$, the block Υ_{ii} includes $\mathcal{B}_{ia}$. Therefore, $\sum_{j \in \mathcal{V}^a} \Upsilon_{ij} \succeq \mathbf{0}$ for all $i \in \mathcal{V}^a$, specifically

$$\sum_{j \in \mathcal{V}^a} \Upsilon_{ij} = \mathbf{0} \text{ if } a \notin \mathcal{N}_i, \text{ and } \quad \sum_{j \in \mathcal{V}^a} \Upsilon_{ij} = \mathcal{B}_{ia} \succ \mathbf{0}, \text{ when } a \in \mathcal{N}_i. \quad (3.42)$$

Since $\mathcal{G}$ is connected, $a \in \mathcal{N}_i$ for at least one robot $i \in \mathcal{V}^a$. Now consider a proper subset $\mathcal{J} \subsetneq \mathcal{V}^a$. Note that for each $i \in \mathcal{J} \subsetneq \mathcal{V}^a$,

$$\sum_{j \in \mathcal{J}} \Upsilon_{ij} = \mathbf{0} \text{ if } \mathcal{N}_i \subseteq \mathcal{J}, \text{ and } \quad \sum_{j \in \mathcal{J}} \Upsilon_{ij} = \sum_{j \in \mathcal{N}_i \setminus \mathcal{J}} \mathcal{B}_{ij} \succ \mathbf{0}, \text{ otherwise.}$$

$$(3.43)$$

Since $\mathcal{G}$ is connected, given any proper subset $\mathcal{J} \subsetneq \mathcal{V}^a$ of robots, there is always a robot $i \in \mathcal{J}$ that has at least one neighbor outside $\mathcal{J}$ or that has the anchor a as a neighbor, for which $\sum_{j \in \mathcal{J}} \Upsilon_{ij} \succ \mathbf{0}$. Therefore Lemma 4 holds, and by applying Theorem 9 taking $u_2, \ldots, u_n = 1$ we conclude that matrix $\Upsilon \in M_{n-1}^p$. Since Υ is symmetric, then $\Upsilon + \Upsilon^T \in M_{n-1}^p$, and by [17, Theorem 4.7] we conclude that $\rho(D^{-1}N) < 1$.   □

Corollary 4 *Let $\mathcal{G}$ be connected, $\Sigma_{\mathbf{z}_1}, \ldots, \Sigma_{\mathbf{z}_m}$ be the covariance matrices associated to m independent $p-$dimensional measurements, and $\Sigma_{\mathbf{z}}$ be their associated block-diagonal covariance matrix as in Eq. (3.33). Consider that each robot $i \in \mathcal{V}$ executes the Algorithm 3 to update its variable $\hat{\mathbf{x}}_i^a(t)$. Then, for all $i \in \mathcal{V}$,*

$$\lim_{t \to \infty} \hat{\mathbf{x}}_i^a(t) = \hat{\mathbf{x}}_i^a, \quad (3.44)$$

converges to the anchor-based centralized solution $\hat{\mathbf{x}}_i^a$ given by Eq. (3.34).   □

---

[1] $A \succ B$ ($A \succeq B$) represent that matrix $A - B$ is positive-definite (positive-semidefinite). Equivalently, $\prec, \preceq$ are used for negative-definite and negative-semidefinite matrices.

3.4.2 Centralized Centroid-Based Position Estimation

The accuracy of the estimated states $\hat{\mathbf{x}}_{\mathscr{V}}^{a}$, $\Sigma_{\hat{\mathbf{x}}_{\mathscr{V}}^{a}}$ in anchor-based approaches depend on the selected anchor a. Instead of that it is more interesting to compute the states of the robots $\hat{\mathbf{x}}_{\mathscr{V}}^{cen}$, $\Sigma_{\hat{\mathbf{x}}_{\mathscr{V}}^{cen}}$ relative to the *centroid* given by the average of the states,

$$\hat{\mathbf{x}}_{\mathscr{V}}^{cen} = (\mathbf{I} - H_{cen})\,\hat{\mathbf{x}}_{\mathscr{V}}^{a}, \quad \Sigma_{\hat{\mathbf{x}}_{\mathscr{V}}^{cen}} = (\mathbf{I} - H_{cen})\,\Sigma_{\hat{\mathbf{x}}_{\mathscr{V}}^{a}}\,(\mathbf{I} - H_{cen})^{T}, \qquad (3.45)$$

$$\text{where } H_{cen} = (\mathbf{1}_n \otimes \mathbf{I}_p)\,(\mathbf{1}_n \otimes \mathbf{I}_p)^{T}/n.$$

The value of this representation is that the states of the robots $\hat{\mathbf{x}}_{\mathscr{V}}^{cen}$, $\Sigma_{\hat{\mathbf{x}}_{\mathscr{V}}^{cen}}$ with respect to the centroid are the same regardless of the anchor robot, i.e., the centroid solution is unique. Additionally, as the following result shows, it produces more accurate estimates than the ones provided by any anchor selection. We compare the block-traces[2] blkTr of their covariance matrices [6].

Proposition 5 *The covariance matrices of the centroid-based* $\Sigma_{\hat{\mathbf{x}}_{\mathscr{V}}^{cen}}$ *and anchor-based* $\Sigma_{\hat{\mathbf{x}}_{\mathscr{V}}^{a}}$ *estimates satisfy, for all anchors* $a \in \mathscr{V}$,

$$\text{blkTr}\left(\Sigma_{\hat{\mathbf{x}}_{\mathscr{V}}^{cen}}\right) \preceq \text{blkTr}\left(\Sigma_{\hat{\mathbf{x}}_{\mathscr{V}}^{a}}\right), \qquad \text{Tr}\left(\Sigma_{\hat{\mathbf{x}}_{\mathscr{V}}^{cen}}\right) \leq \text{Tr}\left(\Sigma_{\hat{\mathbf{x}}_{\mathscr{V}}^{a}}\right). \qquad (3.46)$$

Proof Let P_{ij} and Q_{ij} be the $p \times p$ blocks of, respectively, the anchor and the centroid-based covariances, $\Sigma_{\hat{\mathbf{x}}_{\mathscr{V}}^{a}} = [P_{ij}]$, $\Sigma_{\hat{\mathbf{x}}_{\mathscr{V}}^{cen}} = [Q_{ij}]$ with $i, j \in \mathscr{V}$. The block-trace of the anchor-based covariance matrix is

$$\text{blkTr}\left(\Sigma_{\hat{\mathbf{x}}_{\mathscr{V}}^{a}}\right) = \sum_{i=1}^{n} P_{ii}. \qquad (3.47)$$

Considering Eq. (3.45), each block in the main diagonal of the centroid-based $\Sigma_{\hat{\mathbf{x}}_{\mathscr{V}}^{cen}}$ covariance matrix is given by

$$Q_{ii} = P_{ii} - \frac{1}{n}\sum_{j=1}^{n}(P_{ij} + P_{ji}) + \frac{1}{n^2}\sum_{j=1}^{n}\sum_{j'=1}^{n}P_{jj'}, \qquad (3.48)$$

for $i \in \mathscr{V}$, and thus its block-trace is

$$\text{blkTr}\left(\Sigma_{\hat{\mathbf{x}}_{\mathscr{V}}^{cen}}\right) = \sum_{i=1}^{n}Q_{ii} = \sum_{i=1}^{n}P_{ii} - \frac{1}{n}\sum_{i=1}^{n}\sum_{j=1}^{n}P_{ij}$$

$$= \text{blkTr}\left(\Sigma_{\hat{\mathbf{x}}_{\mathscr{V}}^{a}}\right) - (\mathbf{1}_n \otimes \mathbf{I}_p)^{T}\,\Sigma_{\hat{\mathbf{x}}_{\mathscr{V}}^{a}}\,(\mathbf{1}_n \otimes \mathbf{I}_p)/n. \qquad (3.49)$$

---

[2]The block-trace of a matrix defined by blocks $P = [P_{ij}]$ with $i, j \in \{1, \ldots, n\}$ is the sum of its diagonal blocks, $\text{blkTr}(P) = \sum_{i=1}^{n} P_{ii}$.

Since $\Sigma_{\hat{\mathbf{x}}_{\mathcal{V}}^a}$ is symmetric and positive-semidefinite, then $(\mathbf{1}_n \otimes \mathbf{I}_p)^T \Sigma_{\hat{\mathbf{x}}_{\mathcal{V}}^a} (\mathbf{1}_n \otimes \mathbf{I}_p) \succeq \mathbf{0}$, and thus $\text{blkTr}\left(\Sigma_{\hat{\mathbf{x}}_{\mathcal{V}}^{cen}}\right) - \text{blkTr}\left(\Sigma_{\hat{\mathbf{x}}_{\mathcal{V}}^a}\right) \preceq \mathbf{0}$, as in Eq. (3.46). Observe that the trace of the block-trace of a matrix A is equal to its trace, $\text{Tr}(\text{blkTr}(A)) = \text{Tr}(A)$. Since $\text{blkTr}\left(\Sigma_{\hat{\mathbf{x}}_{\mathcal{V}}^{cen}}\right) - \text{blkTr}\left(\Sigma_{\hat{\mathbf{x}}_{\mathcal{V}}^a}\right) \preceq \mathbf{0}$, the elements in the main diagonal of $\text{blkTr}\left(\Sigma_{\hat{\mathbf{x}}_{\mathcal{V}}^{cen}}\right)$ are smaller than or equal to the ones in the main diagonal of $\text{blkTr}\left(\Sigma_{\hat{\mathbf{x}}_{\mathcal{V}}^a}\right)$ so that

$$\text{Tr}(\Sigma_{\hat{\mathbf{x}}_{\mathcal{V}}^{cen}}) = \text{Tr}(\text{blkTr}(\Sigma_{\hat{\mathbf{x}}_{\mathcal{V}}^{cen}})) \leq \text{Tr}(\text{blkTr}(\Sigma_{\hat{\mathbf{x}}_{\mathcal{V}}^a})) = \text{Tr}(\Sigma_{\hat{\mathbf{x}}_{\mathcal{V}}^a}).$$

$\square$

In particular, from Eq. (3.49), $\text{Tr}(\Sigma_{\hat{\mathbf{x}}_{\mathcal{V}}^a}) - \text{Tr}(\Sigma_{\hat{\mathbf{x}}_{\mathcal{V}}^{cen}}) = \frac{1}{n}\sum_{i=1}^{n}\sum_{j=1}^{n}\text{Tr}(P_{ij})$. Note that the previous result holds when the anchor state $\hat{\mathbf{x}}_a^a$ is set to a general value, not necessarily $\mathbf{0}$. It also holds when there is more than one anchor. Consider that the first k robots are anchors. In this case, matrix $\Sigma_{\hat{\mathbf{x}}_{\mathcal{V}}^a} = [P_{ij}]$ has its blocks $P_{ij} = \mathbf{0}$ for $i, j \in \{1, \ldots, k\}$, and Eq. (3.49) gives $\text{blkTr}(\Sigma_{\hat{\mathbf{x}}_{\mathcal{V}}^{cen}}) = \text{blkTr}(\Sigma_{\hat{\mathbf{x}}_{\mathcal{V}}^a}) - \sum_{i=k+1}^{n}\sum_{j=k+1}^{n} P_{ij}/n$, where $\sum_{i=k+1}^{n}\sum_{j=k+1}^{n} P_{ij}/n \succeq \mathbf{0}$.

We propose an algorithm that allows each robot $i \in \mathcal{V}$ to compute its state $\hat{\mathbf{x}}_i^{cen}$ with respect to the centroid in a distributed fashion, where $\hat{\mathbf{x}}_{\mathcal{V}}^{cen}$ is given in Eq. (3.45), $\hat{\mathbf{x}}_{\mathcal{V}}^{cen} = ((\hat{\mathbf{x}}_1^{cen})^T, \ldots, (\hat{\mathbf{x}}_n^{cen})^T)^T$. These states sum up to zero, $\hat{\mathbf{x}}_1^{cen} + \cdots + \hat{\mathbf{x}}_n^{cen} = \mathbf{0}$, since $(\mathbf{1}_n \otimes \mathbf{I}_p)(\mathbf{I} - H_{cen}) = \mathbf{0}$, and for neighboring robots i and j satisfy $\hat{\mathbf{x}}_i^{cen} = \hat{\mathbf{x}}_j^{cen} - \hat{\mathbf{x}}_j^a + \hat{\mathbf{x}}_i^a$. Thus, a straightforward solution would consist of firstly computing the anchor-based states of the robots $\hat{\mathbf{x}}_{\mathcal{V}}^a = ((\hat{\mathbf{x}}_1^a)^T, \ldots, (\hat{\mathbf{x}}_n^a)^T)^T$, and in a second phase initializing the robots' variables so that they sum up to zero, $\hat{\mathbf{x}}_i^{cen}(0) = \mathbf{0}$, for $i \in \mathcal{V}$, and updating them at each step t with an averaging algorithm that conserves the sum:

$$\hat{\mathbf{x}}_i^{cen}(t+1) = \sum_{j \in \mathcal{N}_i \cup \{i\}} \mathcal{W}_{i,j}(\hat{\mathbf{x}}_j^{cen}(t) - \hat{\mathbf{x}}_j^a + \hat{\mathbf{x}}_i^a), \qquad (3.50)$$

for $i \in \mathcal{V}$, where $\mathcal{W} = [\mathcal{W}_{i,j}]$ is a doubly stochastic weight matrix such that $\mathcal{W}_{i,j} > 0$ if $(i, j) \in \mathcal{E}$ and $\mathcal{W}_{i,j} = 0$ when $j \notin \mathcal{N}_i$. Besides, $\mathcal{W}_{i,i} \in [\alpha, 1]$, $\mathcal{W}_{i,j} \in \{0\} \cup [\alpha, 1]$ for all $i, j \in \mathcal{V}$, for some $\alpha \in (0, 1]$. More information about averaging algorithms can be found in Appendix A and at [9, 26, 37]. The term $-\hat{\mathbf{x}}_j^a + \hat{\mathbf{x}}_i^a$ is the relative measurement $\mathbf{z}_e$ with $e = (j, i)$ for noise free scenarios, and the optimal or corrected measurement [28] $\hat{\mathbf{z}}_e$ for the noisy case, $\hat{\mathbf{z}} = (\mathscr{A} \otimes \mathbf{I}_p)^T \hat{\mathbf{x}}_{\mathcal{V}}^a$, with $\hat{\mathbf{z}} = ((\hat{\mathbf{z}}_1)^T, \ldots, (\hat{\mathbf{z}}_m)^T)^T$. In what follows we propose an algorithm where, at each iteration t, (3.50) is executed not on the exact $\hat{\mathbf{x}}_i^a$, $\hat{\mathbf{x}}_j^a$, but on the most recent estimates $\hat{\mathbf{x}}_i^a(t)$, $\hat{\mathbf{x}}_j^a(t)$ obtained with Algorithm 3.

3.4.3 Distributed Centroid-Based Position Estimation

Now we study a distributed localization algorithm for estimating the position the robots relative to the centroid.

Algorithm 4 Let each robot $i \in \mathcal{V}$ have an estimate of its own state relative to the centroid, $\hat{\mathbf{x}}_i^{cen}(t) \in \mathbb{R}^p$, initialized at $t = 0$ with $\hat{\mathbf{x}}_i^{cen}(0) = \mathbf{0}$. At each time step t, each robot $i \in \mathcal{V}$ updates $\hat{\mathbf{x}}_i^{cen}(t)$ with

$$\hat{\mathbf{x}}_i^{cen}(t+1) = \sum_{j \in \mathcal{N}_i \cup \{i\}} \mathcal{W}_{i,j}(\hat{\mathbf{x}}_j^{cen}(t) + \hat{\mathbf{x}}_i^a(t) - \hat{\mathbf{x}}_j^a(t)), \qquad (3.51)$$

where $\hat{\mathbf{x}}_i^a(t)$, $\hat{\mathbf{x}}_j^a(t)$ are the most recent estimates that robots i and j have at iteration t of the variables in Algorithm 3 and $\mathcal{W}_{i,j}$ are the Metropolis weights as defined in Eq. (A.3) in Appendix A.

Theorem 7 *Let all the robots $i \in \mathcal{V}$ execute the Algorithm 4 and let $\mathcal{G}$ be connected. Then, the estimated states $\hat{\mathbf{x}}_i^{cen}(t)$ at each robot $i \in \mathcal{V}$ asymptotically converge to the state of i relative to the centroid $\hat{\mathbf{x}}_i^{cen}$ given by Eq. (3.45),*

$$\lim_{t \to \infty} \hat{\mathbf{x}}_i^{cen}(t) = \hat{\mathbf{x}}_i^{cen}. \qquad (3.52)$$

Let $\mathbf{e}_{cen}(t) = \left[(\hat{\mathbf{x}}_1^{cen}(t) - \hat{\mathbf{x}}_1^{cen})^T, \ldots, (\hat{\mathbf{x}}_n^{cen}(t) - \hat{\mathbf{x}}_n^{cen})^T \right]^T$ be the error vector containing the estimation errors of the n robots at iteration t. For fixed communication graphs $\mathcal{G}$, the norm of the error vector after t iterations of Algorithm 4 satisfies

$$||\mathbf{e}_{cen}(t)||_2 \le \lambda_{\text{eff}}^t(\mathcal{W})||\mathbf{e}_{cen}(0)||_2 + 2p(n-1)\sigma_J \lambda_{\text{eff}}^t(\mathcal{W}) \sum_{k=1}^{t} \left(\frac{\rho(J)}{\lambda_{\text{eff}}(\mathcal{W})} \right)^k,$$

$$(3.53)$$

where J is the Jacobi matrix $J = D^{-1}N$, with D and N computed as in Eqs. (3.35)– (3.37), σ_J is a constant that depends on the initial Jacobi error and on J. $\mathcal{W}$ is the Metropolis weight matrix as defined in Eq. (A.3) in Appendix A, and $\mathbf{e}_{cen}(0)$ is the initial error at $t = 0$.

Proof First of all, we derive the expression for the convergence rate in Eq. (3.53). We express Algorithm 4 in terms of the error vectors associated to the centroid $\mathbf{e}_{cen}(t)$ and the anchor-based $\mathbf{e}_a(t) \in \mathbb{R}^{(n-1)p}$ estimation methods (Algorithms 3 and 4),

$$\mathbf{e}_{cen}(t) = \left[(\hat{\mathbf{x}}_1^{cen}(t))^T, \ldots, (\hat{\mathbf{x}}_n^{cen}(t))^T \right]^T - \hat{\mathbf{x}}_{\mathcal{V}}^{cen},$$

with $\hat{\mathbf{x}}_{\mathscr{V}}^{cen} = \left[(\hat{\mathbf{x}}_1^{cen})^T, \dots, (\hat{\mathbf{x}}_n^{cen})^T \right]^T$ given by Eq. (3.45), and

$$\tilde{\mathbf{e}}_a(t) = \left[(\hat{\mathbf{x}}_2^a(t)^T, \dots, \hat{\mathbf{x}}_n^a(t)^T \right]^T - \hat{\mathbf{x}}_{\mathscr{V}a}^a,$$

with $\hat{\mathbf{x}}_{\mathscr{V}a}^a = \left[(\hat{\mathbf{x}}_2^a)^T, \dots, (\hat{\mathbf{x}}_n^a)^T \right]^T$ given by Eq. (3.34), where for simplicity we let the robot $i = 1$ be the anchor a. We let $\mathbf{e}_a(t)$ be $(\mathbf{0}_p^T, \tilde{\mathbf{e}}_a(t)^T)^T$. Recall that $\sum_{j \in \mathcal{N}_i \cup \{i\}} \hat{\mathbf{x}}_i^a(t) = \hat{\mathbf{x}}_i^a(t)$ and that the estimated states relative to the centroid $\hat{\mathbf{x}}_{\mathscr{V}}^{cen}$ are $\hat{\mathbf{x}}_{\mathscr{V}}^{cen} = (\mathbf{I} - H_{cen})\hat{\mathbf{x}}_{\mathscr{V}}^a$ as in Eq. (3.45). Algorithm 4 becomes

$$\mathbf{e}_{cen}(t) = (\mathscr{W} \otimes \mathbf{I}_p)\mathbf{e}_{cen}(t-1) + ((\mathbf{I}_n - \mathscr{W}) \otimes \mathbf{I}_p)\mathbf{e}_a(t-1) + P\hat{\mathbf{x}}_{\mathscr{V}}^a, \quad (3.54)$$

where the term P that is multiplying $\hat{\mathbf{x}}_{\mathscr{V}}^a$ is

$$P = \mathbf{I} - (\mathscr{W} \otimes \mathbf{I}_p) - (\mathbf{I} - (\mathscr{W} \otimes \mathbf{I}_p))(\mathbf{I} - H_{cen}) = (\mathbf{I} - (\mathscr{W} \otimes \mathbf{I}_p))H_{cen}. \quad (3.55)$$

We use the fact that $(\mathscr{W} \otimes \mathbf{I}_p)H_{cen} = H_{cen}$, and the previous expression gives $P = \mathbf{0}$ and Eq. (3.54) becomes

$$\mathbf{e}_{cen}(t) = (\mathscr{W} \otimes \mathbf{I}_p)\mathbf{e}_{cen}(t-1) + ((\mathbf{I}_n - \mathscr{W}) \otimes \mathbf{I}_p)\mathbf{e}_a(t-1)$$

$$= (\mathscr{W} \otimes \mathbf{I}_p)^t \mathbf{e}_{cen}(0) + \sum_{k=0}^{t-1} (\mathscr{W} \otimes \mathbf{I}_p)^{t-k-1} \left((\mathbf{I} - \mathscr{W}) \otimes \mathbf{I}_p \right) \mathbf{e}_a(k). \quad (3.56)$$

Then, the norm of the error $\mathbf{e}_{cen}(t)$ satisfies

$$\|\mathbf{e}_{cen}(t)\|_2 \le \lambda_{\text{eff}}^t(\mathscr{W})\|\mathbf{e}_{cen}(0)\|_2 + 2\sum_{k=0}^{t-1} \lambda_{\text{eff}}^{t-k-1}(\mathscr{W})\|\mathbf{e}_a(k)\|_2, \quad (3.57)$$

where we have used the fact that $\| ((\mathscr{W} - \mathbf{I}) \otimes \mathbf{I}_p) \|_2 \le 2$ since $\mathscr{W}$ is the Metropolis weight matrix given by Eq. (A.3) in Appendix A.

We analyze now the norm of error $\|\mathbf{e}_a(t)\|_2$, which is related to the error vector of the Jacobi algorithm $\tilde{\mathbf{e}}_a(t) \in \mathbb{R}^{(n-1)p}$ by $\mathbf{e}_a(t) = (\mathbf{0}, \tilde{\mathbf{e}}_a^T(t))^T$. Let J be the Jacobi matrix, and $V_J = \left[\mathbf{v}_{p+1}(J), \dots, \mathbf{v}_{np}(J) \right]$ and $\lambda_J = \text{Diag}\left(\lambda_{p+1}(J), \dots, \lambda_{np}(J) \right)$ be its associated eigenvectors and eigenvalues so that $J = V_J \lambda_J V_J^{-1}$, and $\|\mathbf{v}_i(J)\|_2 = 1$. The error vector $\tilde{\mathbf{e}}_a(t)$ evolves according to

$$\tilde{\mathbf{e}}_a(t) = J\tilde{\mathbf{e}}_a(t-1) = J^t \tilde{\mathbf{e}}_a(0). \quad (3.58)$$

For each initial error vector $\tilde{\mathbf{e}}_a(0)$ there exist $\sigma_{p+1}, \ldots, \sigma_{np}$ such that

$$\tilde{\mathbf{e}}_a(0) = \sum_{i=p+1}^{np} \sigma_i \mathbf{v}_i(J),$$

and then the error vector $\tilde{\mathbf{e}}_a(t)$ after t iterations of the Jacobi algorithm given by Eq. (3.58) can be expressed as

$$\tilde{\mathbf{e}}_a(t) = V_J \lambda_J^t V_J^{-1} V_J \left[\sigma_{p+1}, \ldots, \sigma_{np} \right]^T = \sum_{i=p+1}^{np} \sigma_i \mathbf{v}_i(J) \lambda_i^t(J).$$

Let $\sigma_J = \max_{i=p+1}^{np} |\sigma_i|$, and $\rho(J) = \max_{i=p+1}^{np} |\lambda_i(J)|$. For all $t \geq 0$, the norm of the error vector $||\tilde{\mathbf{e}}_a(t)||_2$ satisfies

$$||\mathbf{e}_a(t)||_2 = ||\tilde{\mathbf{e}}_a(t)||_2 \leq p(n-1)\sigma_J \rho^t(J). \tag{3.59}$$

Linking this with Eq. (3.57) gives that the convergence rate is

$$||\mathbf{e}_{cen}(t)||_2 \leq \lambda_{\text{eff}}^t(\mathscr{W})||\mathbf{e}_{cen}(0)||_2 + 2p(n-1)\sigma_J \sum_{k=0}^{t-1} \lambda_{\text{eff}}^{t-k-1}(\mathscr{W})\rho^k(J), \tag{3.60}$$

as in Eq. (3.53).

Now we prove the asymptotical convergence to the centroid (3.52). If both the Jacobi and the general algorithm have the same convergence rate, $\rho(J) = \lambda_{\text{eff}}(\mathscr{W})$, then Eq. (3.60) gives

$$||\mathbf{e}_{cen}(t)||_2 \leq \lambda_{\text{eff}}^t(\mathscr{W})||\mathbf{e}_{cen}(0)||_2 + 2p(n-1)\sigma_J \lambda_{\text{eff}}^{t-1}(\mathscr{W})t, \tag{3.61}$$

whereas for $\rho(J) \neq \lambda_{\text{eff}}(\mathscr{W})$, it gives

$$||\mathbf{e}_{cen}(t)||_2 \leq \lambda_{\text{eff}}^t(\mathscr{W})||\mathbf{e}_{cen}(0)||_2 + \frac{2p(n-1)\sigma_J}{\rho(J) - \lambda_{\text{eff}}(\mathscr{W})}(\rho^t(J) - \lambda_{\text{eff}}^t(\mathscr{W})). \tag{3.62}$$

Note that $\lambda_{\text{eff}}(\mathscr{W}) < 1$ for connected graphs $\mathscr{G}$. Then, the term $\lambda_{\text{eff}}^t(\mathscr{W})||\mathbf{e}_{cen}(0)||_2$ in Eqs. (3.61) and (3.62) exponentially tends to zero as $t \to \infty$ regardless of the initial error $\mathbf{e}_{cen}(0)$. For the case $\rho(J) = \lambda_{\text{eff}}(\mathscr{W})$, the term $\lambda_{\text{eff}}^t(\mathscr{W})t$ in Eq. (3.61) is decreasing for $t \geq \frac{\lambda_{\text{eff}}(\mathscr{W})}{1 - \lambda_{\text{eff}}(\mathscr{W})}$ and thus it tends to zero as $t \to \infty$. For $\rho(J) \neq \lambda_{\text{eff}}(\mathscr{W})$, the term $(\rho^t(J) - \lambda_{\text{eff}}^t(\mathscr{W}))$ in Eq. (3.62) asymptotically tends to zero since $\lambda_{\text{eff}}(\mathscr{W})$ is less than 1, and as stated by Theorem 6, $\rho(J) < 1$. Therefore, $\lim_{t \to \infty} ||\mathbf{e}_{cen}(t)||_2 = 0$, where $||\mathbf{e}_{cen}(t)||_2 = 0$ iff $\mathbf{e}_{cen}(t) = 0$, what concludes the proof. □

3.5 Simulations

Planar Localization from Noisy Measurements

A set of simulations have been carried out to show the performance of the method for planar localization from noisy measurements, and to compare the results of the centralized (Sect. 3.3.1) and the distributed (Sect. 3.3.2) approaches.

First, a team of $n = 20$ robots are placed in a ring of radius 4 m with their orientations randomly generated within $\pm\frac{\pi}{2}$ (Fig. 3.1a). Each robot measures the relative pose of the next robot, with noises in the x- and y- coordinates of 6 cm standard deviation, and of 1 degree for the orientation. The robots execute the proposed method to compute their pose with respect to the anchor node $R1$. The experiment is repeated 100 times and the average results can be seen in Table 3.1. The first rows show the solution computed by the localization algorithm in Sect. 3.3.1, and the next rows compare the distributed implementation of the algorithm (Sect. 3.3.2) against the results obtained by the centralized algorithm in Sect. 3.3.1. We use the flagged initialization [4] that is known to produce fast convergence results. The convergence speeds during the first and the third phases depend on the values of respectively $\rho(C^{-1}D)$ in (3.18) and $\rho(\mathbf{I} - M^{-1}\Upsilon_{\hat{\mathbf{q}}^a_{\mathscr{Y}^a}})$ in (3.31), which here are close to one (slow convergence). The second phase is always executed in a single iteration (it does not have any convergence speed associated). After executing the first phase for $t = 50$ iterations, the obtained $\bar{\theta}^a_{\mathscr{Y}}$ still differs from the centralized solution $\tilde{\theta}^a_{\mathscr{Y}}$ by around $0.16°$. If we increase the number of iterations we obtain better approximations that differ only by 0.01 ($t = 100$) and $8.5e − 05$ ($t = 200$) degrees. The next three rows show the results after executing the second phase followed by 200 iterations of the third phase. Since the second and third phases have been executed using $\bar{\theta}^a_{\mathscr{Y}}$ instead of $\tilde{\theta}^a_{\mathscr{Y}}$, the final results also differ. For the case $t = 200$ (third column), the difference between the pose estimated by the distributed and centralized approaches

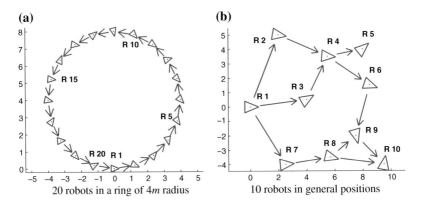

(a)
20 robots in a ring of 4*m* radius

(b)
10 robots in general positions

Fig. 3.1 Scenarios tested. Each robot (*triangles*) measures the relative pose of other team members (*outgoing arrows*). Robots connected by an arrow can exchange data

Table 3.1 Results for the scenario in Fig. 3.1a

Localization results versus ground truth

	Max error	Average standard deviation
Orientation phase 1	3.38°	1.87°
x-coordinate phase 3	27.85 cm	13.45 cm
y-coordinate phase 3	24.33 cm	12.31 cm
Orientation phase 3	4.03°	1.66°

Distributed implementation (flagged-initialization)

$\rho(C^{-1}D)$	0.99		
$\rho(\mathbf{I} - M^{-1}\Upsilon_{\hat{\mathbf{q}}^a_{\gamma a}})$	0.99		
Max error	$t = 50$	$t = 100$	$t = 200$
Orientation phase 1	0.16°	0.01°	$8.5e - 05°$
x-coordinate phase 3	1.74 cm	1.64 cm	1.64 cm
y-coordinate phase 3	0.84 cm	0.49 cm	0.48 cm
Orientation phase 3	0.29°	0.12°	0.11°

is small (1.64 and 0.48 cm for the x- and y- coordinates, and 0.11 degrees for the orientation), and similar results are obtained for $t = 100$.

Other simulation with 10 robots placed as in Fig. 3.1b has been carried out. If there is an arrow from robot i into j, then robot i measures the relative pose of robot j, with additive noises of 2.5 degrees of standard deviation for the orientation, and of 5 % d and 0.7 % d standard deviation for respectively the x and y-coordinates, where d is the distance between the robots. The robots execute the distributed algorithm during the phase 1 to compute their orientations with respect to the anchor node R1 (Fig. 3.2a), obtaining estimates (blue) very close to the ground truth data (red). They execute phase 2 to express the relative position measurements in the reference

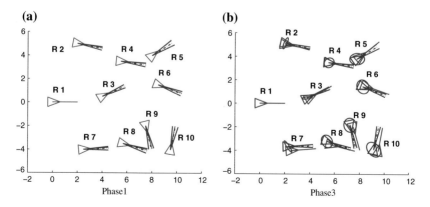

Fig. 3.2 The robots estimate their poses (*blue dashed*) relative to the anchor R1 for the experiment in Fig. 3.1b. The ground truth data (*red solid*) and the covariances computed by the centralized approach (*blue solid*) are also displayed

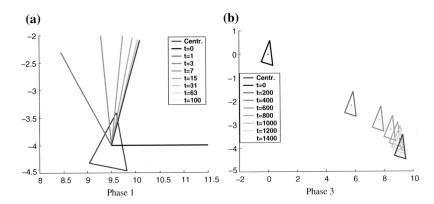

Fig. 3.3 Detail of phases 1 and 3 of the proposed strategy. The estimates of robot R10 (*gray*) successively approach the centralized solution (*blue*)

frame of the anchor node. Finally, they execute the phase 3 to obtain both, their positions and orientations relative to the anchor node (Fig. 3.2b). Figure 3.3 shows a detail of the iterations during phases 1 and 3. Although the convergence was previously proved only for graphs with low connectivity (ring or string graphs), in the experiments with general communication graphs the algorithm has been found to converge as well.

Centroid-Based Noisy Position Localization

We study the performance of the algorithm presented in Sect. 3.4 in a multi-robot localization scenario (Fig. 3.4) with $n = 20$ robots (black circles) that get noisy measurements (gray crosses and ellipses) of the position of robots which are closer than

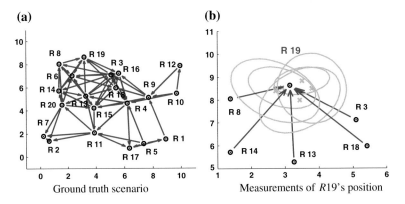

Fig. 3.4 **a** 20 robots (*black circles*) are placed randomly in a region of 10×10 m. There is an edge $e = (i, j) \in \mathcal{E}$ (*red arrows*) between pairs of robots that are closer than 4 m. **b** Each robot i has a noisy measurement $\mathbf{z}_e$ (*gray crosses* and *ellipses*) of the relative position of its out-neighbors j, with $e = (i, j)$. The noises are proportional to the distance between the robots

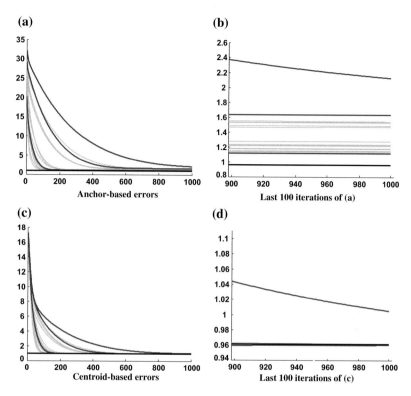

(a)

Anchor-based errors

(b)

Last 100 iterations of (a)

(c)

Centroid-based errors

(d)

Last 100 iterations of (c)

Fig. 3.5 The experiment in Fig. 3.4 is generated 100 times with the same noise levels but different noise values. We display the average norm of the error with the difference between the estimates and the ground truth for the 100 different experiments. **a** Results of Algorithm 3 when each robot $i \in \mathscr{V}$ is used as the anchor (*gray lines*). The special cases that the anchor is $R1$, $R3$ and $R12$ are shown in *blue*. The *black line* is the asymptotic error reached with the centroid-based estimation. **b** Detail of iterations 900–1000 in (*a*). **c** Results of Algorithm 4 using all the possible anchors. **d** Detail of iterations 900–1000 in (*c*)

4 m. We analyze the states estimated by the n robots along 1000 iterations of the proposed algorithm (Fig. 3.5). Robots initialize their states $\hat{\mathbf{x}}_i^a(t)$, $\hat{\mathbf{x}}_i^{cen}(t)$ with zeros and execute Algorithms 3 and 4. We generate specific noises as the ones in Fig. 3.4 for 100 different samples. For each of them, we record the norm of the error vector containing the difference between the estimates at the n robots and the ground-truth positions at each iteration t. In Fig. 3.5a we show the results of Algorithm 3 when each robot $i \in \mathscr{V}$ is used as the anchor (gray lines). The most and least precise anchor-based results, which are obtained for respectively $R3$ and $R12$, are shown in blue. The results for robot $R1$, which is conventionally used as the anchor, are displayed in blue as well. The black line is the asymptotic error reached with the centroid-based estimation method. As it can be seen, the errors reached with the anchor-based solutions are greater than the ones associated to the centroid. This is even more evident in Fig. 3.5b, which shows the last 100 iterations in Fig. 3.5a. In

Fig. 3.5c we show the equivalent errors for the centroid-based estimation algorithm (Algorithm 4), using all the possible anchors for Algorithm 3. Here, in all cases the error estimates (gray lines) converge to the asymptotic error of the centroid-based estimation method (black line).

3.6 Closure

Along this chapter, the problem of network localization has been studied for different scenarios: the estimation of the planar localization with respect to an anchor from noisy relative measurements, and the estimation of higher dimension positions with respect to the, simultaneously computed, centroid of the network using also noisy measurements. We have analyzed distributed strategies that allow the robots to agree on a common global frame, and to compute their poses relative to the global frame. The presented algorithms exclusively rely on local computations and data exchange with direct neighbors. Besides, they only require the robots to maintain their own estimated poses relative to the common frame. Thus, the memory load of the algorithm is low compared to methods where each robot must also estimate the positions or poses of any other robot. We have discussed the performance of the planar pose localization algorithm relative to an anchor node, for ring or string topologies. The centroid-based position localization method has been studied to produce more accurate results than any anchor-based solution. Besides, in the experiments we have shown that it converges faster than the anchor-based solutions.

References

1. B.D.O. Anderson, I. Shames, G. Mao, B. Fidan, Formal theory of noisy sensor network localization. SIAM J. Discret. Math. **24**(2), 684–698 (2010)
2. R. Aragues, L. Carlone, G. Calafiore, C. Sagues, Multi agent localization from noisy relative pose measurements, in *IEEE International Conference on Robotics and Automation* , Shanghai, China, May 2011, pp. 364–369
3. R. Aragues, L. Carlone, C. Sagues, G. Calafiore, Distributed centroid estimation from noisy relative measurements. Syst. Control Lett. **61**(1), 773–779 (2012)
4. P. Barooah, J. Hespanha, Distributed estimation from relative measurements in sensor networks, in *International Conference on Intelligent Sensing and Information Processing*, Chennai, India, January 2005, pp. 88–93
5. P. Barooah, J. Hespanha, Estimation on graphs from relative measurements. IEEE Control Syst. Mag. **27**(4), 57–74 (2007)
6. P. Barooah, J. Hespanha, Error scaling laws for linear optimal estimation from relative measurements. IEEE Trans. Inf. Theory **55**(12), 5661–5673 (2009)
7. D.J. Bennet, C.R. McInnes, Distributed control of multi-robot systems using bifurcating potential fields. Robot. Auton. Syst. **58**(3), 256–264 (2010)
8. D.P. Bertsekas, J.N. Tsitsiklis, *Parallel and Distributed Computation: Numerical Methods* (Athena Scientific, 1997)

9. F. Bullo, J. Cortes, S. Martinez, *Distributed Control of Robotic Networks*. Applied Mathematics Series (Princeton University Press, 2009). Electronically available at http://coordinationbook. info

10. G. Calafiore, L. Carlone, M. Wei, A distributed gauss-newton approach for range-based localization of multi agent formations, in *IEEE Multi-Conference on Systems and Control*, Yokohama, Japan, September 2010, pp. 1152–1157

11. G. Calafiore, L. Carlone, M. Wei, A distributed gradient method for localization of formations using relative range measurements, in *IEEE Multi-Conference on Systems and Control*, Yokohama, Japan, September 2010, pp. 1146–1151

12. L. Carlone, R. Aragues, J.A. Castellanos, B. Bona, A first-order solution to simultaneous localization and mapping with graphical models, in *IEEE International Conference on Robotics and Automation*, Shanghai, China, May 2011, pp. 1764–1771

13. L. Carlone, R. Aragues, J.A. Castellanos, B. Bona, A linear approximation for graphbased simultaneous localization and mapping, in *Robotics: Science and Systems*, Los Angeles, CA, USA, June 2011

14. S. Carpin, Fast and accurate map merging for multi-robot systems. Auton. Robot. **25**(3), 305–316 (2008)

15. S. Carpin, A. Birk, V. Jucikas, On map merging. Robot. Auton. Syst. **53**(1), 1–14 (2005)

16. J. Cortes, Global and robust formation-shape stabilization of relative sensing networks. Automatica **45**(12), 2754–2762 (2009)

17. L. Elsner, V. Mehrmann, Convergence of block iterative methods for linear systems arising in the numerical solution of Euler equations. Numerische Mathematik **59**(1), 541–559 (1991)

18. J.A. Fax, R.M. Murray, Information flow and cooperative control of vehicle formations. IEEE Trans. Autom. Control **49**(9), 1465–1476 (2004)

19. M. Franceschelli, A. Gasparri, On agreement problems with gossip algorithms in absence of common reference frames, in *IEEE International Conference on Robotics and Automation*, Anchorage, USA, May 2010, pp. 4481–4486

20. J.J. Guerrero, A.C. Murillo, C. Sagues, Localization and matching using the planar trifocal tensor with bearing-only data. IEEE Trans. Robot. **24**(2), 494–501 (2008)

21. M. Ji, M. Egerstedt, Distributed coordination control of multiagent systems while preserving connectedness. IEEE Trans. Robot. **23**(4), 693–703 (2007)

22. J. Knuth, P. Barooah, Distributed collaborative localization of multiple vehicles from relative pose measurements, in *Allerton Conference on Communications, Control and Computing*, Urbana-Champaign, USA, October 2009, pp. 314–321

23. L. Kolotilina, Bounds for eigenvalues of symmetric block Jacobi scaled matrices. J. Math. Sci. **79**(3), 1043–1047 (1996)

24. G. Lafferriere, A. Williams, J. Caughman, J.J.P. Veerman, Decentralized control of vehicle formations. Syst. Control Lett. **54**(9), 899–910 (2005)

25. N. Mostagh, A. Jadbabaie, Distributed geodesic control laws for flocking of nonholonomic agents. IEEE Trans. Autom. Control **52**(4), 681–686 (2007)

26. W. Ren, R.W. Beard, *Distributed Consensus in Multi-vehicle Cooperative Control Communications and Control Engineering* (Springer, London, 2008)

27. S.I. Roumeliotis, G.A. Bekey, Distributed multirobot localization. IEEE Trans. Robot. Autom. **18**(5), 781–795 (2002)

28. W.J. Russell, D. Klein, J.P. Hespanha, Optimal estimation on the graph cycle space, in *American Control Conference*, Baltimore, June 2010, pp. 1918–1924

29. C. Sagues, A.C. Murillo, J.J. Guerrero, T. Goedemé, T. Tuytelaars, L. Van Gool, Localization with omnidirectional images using the 1D radial trifocal tensor, in *IEEE International Conference on Robotics and Automation*, Orlando, May 2006, pp. 551–556

30. A. Sarlette, R. Sepulchre, N.E. Leonard, Autonomous rigid body attitude synchronization. Automatica **45**(2), 572–577 (2008)

31. A. Savvides, W.L. Garber, R.L. Moses, M.B. Srivastava, An analysis of error inducing parameters in multihop sensor node localization. IEEE Trans. Mob. Comput. **4**(6), 567–577 (2005)

32. I. Skrjanc, G. Klancar, Optimal cooperative collision avoidance between multiple robots based on Bernstein-Bzier curves. Robot. Auton. Syst. **58**(1), 1–9 (2010)
33. S. Thrun, Y. Liu, Multi-robot SLAM with sparse extended information filters, in *International Symposium of Robotics Research*, Sienna, Italy, October 2003, pp. 254–266
34. N. Trawny, S.I. Roumeliotis, G.B. Giannakis, Cooperative multi-robot localization under communication constraints, in *IEEE International Conference on Robotics and Automation*, Kobe, Japan, May 2009, pp. 4394–4400
35. N. Trawny, X.S. Zhou, K.X. Zhou, S.I. Roumeliotis, Inter-robot transformations in 3-d. IEEE Trans. Robot. **26**(2), 226–243 (2010)
36. B. Varghese, G. McKee, A mathematical model, implementation and study of a swarm system. Robot. Auton. Syst. **58**(3), 287–294 (2010)
37. L. Xiao, S. Boyd, S. Lall, A scheme for robust distributed sensor fusion based on average consensus, in *Symposium on Information Processing of Sensor Networks (IPSN)*, Los Angeles, CA, April 2005, pp. 63–70
38. X.S. Zhou, S.I. Roumeliotis, Robot-to-robot relative pose estimation from range measurements. IEEE Trans. Robot. **24**(6), 1379–1393 (2008)

Chapter 4
Map Merging

Abstract This chapter presents a solution for merging feature-based maps in a robotic network with limited communication. We consider a team of robots exploring an unknown environment. Along its operation, each robot observes the environment and builds and maintains its local stochastic map of the visited region. Simultaneously, the robots communicate and build a global map of the environment. The communication between the robots is limited and, at every time instant, each robot can only exchange data with its neighboring robots. This problem has been traditionally addressed using centralized schemes or broadcasting methods. Instead, in this chapter we study a fully distributed approach which is implementable in scenarios with limited communication. This solution does not rely on a particular communication topology and does not require any central node, making the system robust to individual failures. Each robot computes and tracks the global map based on local interactions with its neighbors. Under mild connectivity conditions on the communication graph, the algorithm asymptotically converges to the global map. In addition, we analyze the convergence speed according to the information increase in the local maps. The results are validated through simulations.

Keywords Map merging · Map fusion · Limited communication · Distributed systems · Parallel computation

4.1 Introduction

As stated through the book, there is a great interest in multi-robot perception in an unknown environment where the team operates and individual robots only observe a portion of it. In such situations, it is of interest for each robot to have a representation of the environment beyond its local map. The fusion of the local observations of all the team members leads to a merged map that contains more precise information and more features. In a static map merging scenario, the information fusion is carried out after the exploration. Dynamic solutions, where the information is merged while the robots operate, are more interesting. They enable other multi-robot tasks such as cooperative exploration, navigation, or obstacle avoidance. In this chapter, we study the problem of dynamic map merging, where each robot's communication radius is limited, and hence the communication topology is not complete.

© The Author(s) 2015
R. Aragues et al., *Parallel and Distributed Map Merging and Localization*,
SpringerBriefs in Computer Science, DOI 10.1007/978-3-319-25886-7_4

While multi-robot localization under communication constraints has received some attention [29, 42], most of the existing multi-robot map merging solutions are extensions of the single robot case under centralized schemes, all-to-all communication among the robots, or broadcasting methods. In [41] a single global map is updated by all the robots. Robots search for features in the global map that have been observed by themselves along the exploration. Then, they use these coincident features to compute implicit measurements (the difference between the Cartesian coordinates of equal features must be zero) and use these constrains to update the map. In [18] maps are represented as constraint graphs, where nodes are scans measured from a robot pose and edges represent the difference between pairs of robot poses. Robot-to-robot measurements are used to merge two local maps into a single map. An optimization phase must be carried out in order to transform the constraint graph into a Cartesian map. Reference [14] also represents the global map using a graph. Nodes are local metric maps and edges describe relative positions between adjacent local maps. The map merging process consists of adding an edge between the maps. Global optimization techniques are applied to obtain the global metric map. Reference [52] merges two maps into a single one using robot-to-robot measurements to align the two maps and then detecting duplicated landmarks and imposing the implicit measurement constraints. Particle filters have been generalized to multi-robot systems assuming that the robots broadcast their controls and their observations [22]. The Constrained Local Submap Filter has been extended to the multi-robot case assuming that each robot builds a local submap and broadcasts it, or transmits it to a central node [49]. Methods based on graph maps of laser scans [18, 27, 39, 48] make each robot build a new node and broadcast it. The same solution could be applied for many existing submap approaches [38]. However, in robot network scenarios, distributed approaches are often necessary because of limited communication, switching topologies, link failures, and limited bandwidth.

The previous methods require that each robot has the capability to communicate with all other robots at every time instant or with a central node. Centralized strategies, where a central node compiles all the information from other robots, performs the computations, and propagates the processed information or decisions to all the robots, have several drawbacks. The whole system can fail if the central node fails, leader selection algorithms may be needed, and a (direct or indirect) communication of all robots with the central system may be required. On the other hand, in distributed systems, all robots play the same role, and therefore the computations can be distributed among all the robots. In addition, distributed systems are naturally more robust to individual failures. In distributed scenarios we cannot assume that the robots can communicate with all other robots at every time instant. A more realistic situation is when, at any time instant, robots can communicate only with a limited number of other robots called their neighbors, e.g., robots within a specific distance. These situations can be best modeled using communication graphs, where nodes correspond to the robots and edges represent communication capabilities between them. Additionally, since robots are moving, the topology of the graph may vary along time, given rise to switching topologies, see for instance [9]. We are interested in map merging solutions for robotic systems with range-limited communication, and where the computations are distributed among the robots.

Distributed estimation methods [1, 13, 20, 26, 33, 35, 36, 46] maintain a joint estimate of a system that evolves with time by combining noisy observations taken by the sensor network. Early approaches sum the measurements from the different robots in IF (Information Filter) form. Measurement updates in IF are additive and therefore information coming from different sensors can be fused in any order and at any time. If the network is complete [33], then the resulting estimator is equivalent to the centralized one. In general networks the problems of cyclic updates or double counting information appear when robots sum the same piece of data more than once. The use of the channel filter [20, 46] avoids these problems in networks with a tree structure. The Covariance Intersection method [26] produces consistent but highly conservative estimates in general networks. More recent approaches [1, 13, 35, 36] use distributed consensus filters to average the measurements taken by the robots. The interest of distributed averaging is that the problems of double counting the information and cyclic updates are avoided. They, however, suffer from the delayed data problem that takes place when the robots execute the state prediction without having incorporated all the measurements taken at the current step [12]. For general communication schemes [35], the delayed data problem leads to an approximate KF (Kalman Filter) estimator. An interesting solution is given in [36] but its convergence is proved in the absence of observation and system noises. In the algorithm proposed in [13], authors prove that the robots' estimates are consistent, although these estimates have disagreement. Other algorithms have been proposed that require the previous offline computation of the gains and weights of the algorithm [1]. The main limitation of all the previous works is that they consider linear systems without inputs, and where the evolution of the system is known by all the robots. Here instead we are interested in more general scenarios, without the previous restrictions. We allow each robot to build its map by using system models not necessarily linear or known by the other robots, or where the robot odometry is modeled as an input, among others. A recent work that does not suffer from the previous limitations is given in [30]. Here each robot records its own measurements and odometry, as well as the observations and odometry from any other robot it encounters. Despite being very interesting and going beyond the state of art, that work has the drawback that robots must maintain an unbounded amount of memory, which depends on the time between meetings. Moreover, if a single robot fails or leaves the network, the whole system fails. Other interesting approach that allows the robots to measure both the landmarks positions as well as their own odometry is given by [15]. Each robot has a single representation of the environment that combines its own data and the measurements of its neighbors, being this representation consistent. The main limitation of this work is that the measured information does not go beyond the neighborhood level. Thus, each robot has a better map than as if it was acting on its own. However, it does not have knowledge about the features observed by robots in farther places of the network.

Most of the previous methods have in common that they combine the data acquired by the different robots in the form of raw measurements, and that the local estimate of each robot contains information from the other robots, i.e., local estimates are not independent. Alternatively, information can be processed in the form of local maps, and these local maps can be kept independent by avoiding the introduction of global information into them; this is what we propose here, and it is also the approach followed in [16]. This strategy has the benefit that each robot can produce meaningful representations of the environment, which allows for several high-level data association methods, as the ones discussed in Chap. 2. Not introducing global data in the local maps has the effect of keeping the local maps of different robots independent. Thus, consensus filters can be used without suffering from the previously mentioned problems of delayed data, and double counting information. An advantage of our approach is the natural robustness that results from its distributed implementation.

The consensus filters literature is greatly wide. A review of the most relevant results can be found in [40] and the references therein. Many recent works consider specific variations of the consensus problem to cope with communication delays [44] or stochastic communication noises [31]. Most of the works in distributed consensus address the static case, i.e., consensus is achieved on a value that depends on the initial conditions of the system. Fewer works [11, 19, 37, 43, 51, 53] consider the dynamic case, where nodes measure a variable along time, and the goal is to track the average of this variable. In map merging scenarios, dynamic consensus strategies are more appealing, since the local maps of the robots will change, and it would be desirable to track the global merged map. Several dynamic consensus methods [19, 37, 43] consider continuous-time systems, and thus they are better suited for systems based on the observation of the states of the neighbors, instead of on communicating the states (in our case, the maps). Reference [53] uses discrete-time communication, but it considers that nodes measure a local continuous physical process. On the other hand, [11, 51] track the average of inputs that change in a discrete-time way, using discrete-time communications. Thus, they are better suited to the problem of map merging, where the local maps are modified at discrete time instances. In our previous approach to the dynamic map merging problem [3], we used consensus algorithms [19, 32] that allowed the latest global map to be weighted with a forgetting factor, as the current global map was computed by the robots. This approach has two limitations: first, robots have to be synchronized, i.e., they must initiate every new map merging phase in a coordinated way; and in second place, the method was designed for graphs which remained fixed during a specific merging phase.

In this chapter, we discuss distributed sensor fusion methods which are intended for independent observations acquired by several sensors along time. Instead of observations, we use the information increments of the local maps, i.e., the differences between the local maps at steps k and $k + 1$, expressed in Information Filter form, as inputs to the algorithm. As we discuss, the convergence and unbiased mean properties of the original algorithm remain valid regardless of this modification. An important property that any estimation method should have is *consistency* [15, 23–25], i.e., if the estimates at the robots are not overconfident. In this chapter, we perform a novel and thorough study of the global map estimated by each robot and each step and prove that they are consistent.

4.2 Problem Description

Throughout the chapter we let n be the number of robots. Indices i, j refer to robots, G to the global map, and A to averaged information matrices and vectors. We use $k, k' \in \mathbb{N}$ for time steps. Constants $\mathtt{szr}$ and $\mathtt{szf}$ represent the size of respectively a robot pose and a feature position.[1] We let $\mathbf{I}$ be the identity matrix, and $\mathbf{0}$ be a $n \times n$ matrix with all its elements equal to zero (if a subindex $n_1 \times n_2$ appears, this specifies their dimensions). Given a matrix W, $[W]_{ij}$ denotes its (i, j) entry. $W \succeq V$ ($\preceq$) indicates that matrix $W - V$ is positive- (negative-) semidefinite.

We consider a team of $n \in \mathbb{N}$ robots exploring an unknown environment. There are $m \in \mathbb{N}$ different static features in the environment and we let $\mathbf{x} \in \mathbb{R}^{\mathscr{M}}$ be the vector with their true positions, with $\mathscr{M} = m\,\mathtt{szf}$. Up to the time step k, the latest map of each robot i contains estimates $\hat{\mathbf{x}}_i^k \in \mathbb{R}^{\mathscr{M}_i^k}$ of the positions of the $m_i^k \leq m$ features observed by robot i, where $\mathscr{M}_i^k = m_i^k\,\mathtt{szf}$, with associated covariance matrix $\Sigma_i^k \in \mathbb{R}^{\mathscr{M}_i^k \times \mathscr{M}_i^k}$. Let $H_i^k \in \{0, 1\}^{\mathscr{M}_i^k \times \mathscr{M}}$ be the observation matrix that relates the elements in $\mathbf{x}$ and $\hat{\mathbf{x}}_i^k$; then, the local map of each robot i contains a partial observation of $\mathbf{x}$,

$$\hat{\mathbf{x}}_i^k = H_i^k \mathbf{x} + \mathbf{v}_i^k, \qquad E\left[\mathbf{v}_i^k\right] = \mathbf{0}, \qquad E\left[\mathbf{v}_i^k (\mathbf{v}_i^k)^T\right] = \Sigma_i^k, \qquad (4.1)$$

where $\mathbf{v}_i^k$ is a zero mean noise with covariance matrix Σ_i^k. Up to the time step k, the latest map of each robot i contains as well estimates $\hat{\mathbf{r}}_i^k \in \mathbb{R}^{\mathscr{R}_i^k}$ of r_i^k of the poses of robot i, where $\mathscr{R}_i^k = r_i^k\,\mathtt{szr}$, with associated covariance matrix $R_i^k \in \mathbb{R}^{\mathscr{R}_i^k \times \mathscr{R}_i^k}$.[2] Let $\mathbf{r}_i^k \in \mathbb{R}^{\mathscr{R}_i^k}$ be the true values for these r_i^k poses of robot i up to step k, then

$$\hat{\mathbf{r}}_i^k = \mathbf{r}_i^k + \mathbf{w}_i^k, \qquad\qquad E\left[\mathbf{w}_i^k \left(\mathbf{w}_i^k\right)^T\right] = R_i^k,$$

$$E\left[\mathbf{w}_i^k\right] = \mathbf{0}, \qquad\qquad E\left[\mathbf{w}_i^k \left(\mathbf{v}_i^k\right)^T\right] = S_i^k, \qquad (4.2)$$

where $\mathbf{w}_i^k$ is a zero mean noise with covariance matrix R_i^k, and $S_i^k \in \mathbb{R}^{\mathscr{R}_i^k \times \mathscr{M}_i^k}$ is the cross-covariance between the estimates of the features' positions $\hat{\mathbf{x}}_i^k$ and the robot poses $\hat{\mathbf{r}}_i^k$ in Eqs. (4.1), (4.2). Note that the linear model in Eq. (4.1) refers to the fact that the local maps are an estimate of the features positions; the observation model associated to the sensor used to build the local maps does not need to be linear.

If at step k the information from the n robots was available, e.g., at a central agent, then the global map containing the estimate $\hat{\mathbf{r}}_{G,1}^k, \ldots, \hat{\mathbf{r}}_{G,n}^k$ of the set of poses of each

---

[1] E.g., $\mathtt{szr} = 3$ for planar robot poses (position (x, y) and orientation θ); $\mathtt{szf} = 2$ or $\mathtt{szf} = 3$ for respectively 2D or 3D environments.

[2] E.g., only the last pose ($r_i^k = 1$), the full robot trajectory, or a subset of the trajectory.

robot $\mathbf{r}_1^k, \ldots, \mathbf{r}_n^k$ up to step k, as well as the estimate $\hat{\mathbf{x}}_G^k$ of the positions of the static features $\mathbf{x}$ could be obtained. The local map of each robot i at step k is a partial observation of these elements (Eqs. (4.1), (4.2)),

$$\begin{bmatrix} \hat{\mathbf{r}}_i^k \\ \hat{\mathbf{x}}_i^k \end{bmatrix} = \begin{bmatrix} L_i^k & \mathbf{0} \\ \mathbf{0} & H_i^k \end{bmatrix} \begin{bmatrix} \mathbf{r}_1^k \\ \vdots \\ \mathbf{r}_n^k \\ \mathbf{x} \end{bmatrix} + \begin{bmatrix} \hat{\mathbf{w}}_i^k \\ \hat{\mathbf{v}}_i^k \end{bmatrix},$$

$$\text{where } L_i^k = [\mathbf{0} \ldots \mathbf{0}, \mathbf{I}_{\mathscr{R}_i^k}, \mathbf{0} \ldots \mathbf{0}]. \tag{4.3}$$

We assume that the noises are independent for different robots $i \neq j$ and all $k, k' \in \mathbb{N}$, since every robot has constructed the map based on its own observations, i.e., $E[\mathbf{w}_i^k (\mathbf{w}_j^{k'})^T] = \mathbf{0}$, $E[\mathbf{v}_i^k (\mathbf{v}_j^{k'})^T] = \mathbf{0}$, and $E[\mathbf{w}_i^k (\mathbf{v}_j^{k'})^T] = \mathbf{0}$. Note that since the local map of a robot i at step k is an evolution of its map at any previous step $k' < k$, then the noises $\mathbf{w}_i^k, \mathbf{v}_i^k$, and the noises $\mathbf{w}_i^{k'}, \mathbf{v}_i^{k'}$ are not independent.

Let $Y_i^k \in \mathbb{R}^{\mathcal{M}_G^k \times \mathcal{M}_G^k}$, $\mathbf{y}_i^k \in \mathbb{R}^{\mathcal{M}_G^k}$ be the information matrix and vector of the local map at robot i and step k in IF form, for $i \in \{1, \ldots, n\}$, where $\mathcal{M}_G^k = \mathscr{R}_1^k + \cdots + \mathscr{R}_n^k + \mathcal{M}$,

$$Y_i^k = \begin{bmatrix} L_i^k & \mathbf{0} \\ \mathbf{0} & H_i^k \end{bmatrix}^T \begin{bmatrix} R_i^k & S_i^k \\ (S_i^k)^T & \Sigma_i^k \end{bmatrix}^{-1} \begin{bmatrix} L_i^k & \mathbf{0} \\ \mathbf{0} & H_i^k \end{bmatrix},$$

$$\mathbf{y}_i^k = \begin{bmatrix} L_i^k & \mathbf{0} \\ \mathbf{0} & H_i^k \end{bmatrix}^T \begin{bmatrix} R_i^k & S_i^k \\ (S_i^k)^T & \Sigma_i^k \end{bmatrix}^{-1} \begin{bmatrix} \hat{\mathbf{r}}_i^k \\ \hat{\mathbf{x}}_i^k \end{bmatrix}. \tag{4.4}$$

The mean vector of the global map containing the estimate $\hat{\mathbf{r}}_{G,1}^k, \ldots, \hat{\mathbf{r}}_{G,n}^k$ of the set of poses of each robot $\mathbf{r}_1^k, \ldots, \mathbf{r}_n^k$ up to step k, as well as the estimate $\hat{\mathbf{x}}_G^k$ of the positions of the static features $\mathbf{x}$ is given by,

$$\left(\left(\hat{\mathbf{r}}_{G,1}^k \right)^T, \ldots, \left(\hat{\mathbf{r}}_{G,n}^k \right)^T, \left(\hat{\mathbf{x}}_G^k \right)^T \right)^T = \left(\sum_{i=1}^n Y_i^k \right)^{-1} \sum_{i=1}^n \mathbf{y}_i^k, \tag{4.5}$$

where term $(\sum_{i=1}^n Y_i^k)^{-1}$ is its associated covariance matrix. Merging the maps in IF form is a common practice [45] since the operation is additive, commutative, and associative.

The global map in Eq. (4.5) is different from the one that would be obtained by a centralized multi-robot SLAM, since the local maps in Eq. (4.4) do not include measurements from the other robots. Equation (4.5) computes the minimum-variance unbiased estimate of $\mathbf{r}_1^k, \ldots, \mathbf{r}_n^k, \mathbf{x}$ given *the local maps* (the maximum-likelihood estimate if the local maps are Gaussian), whereas centralized multi-robot SLAM methods estimate $\mathbf{r}_1^k, \ldots, \mathbf{r}_n^k, \mathbf{x}$ given the *measurements and control inputs*. Thus, the accuracy of the global map in Eq. (4.5) depends on the precision of the local maps. The unbiased mean and consistency properties of the global map depend on the local maps having unbiased mean and being consistent. Since we do not include measurements from the other robots, the local maps of different robots remain independent and can be fused by the addition of the information matrices and vectors as in Eq. (4.5).

Now consider the next time step $k + 1$. Robots have kept on exploring and some of the robot maps have changed. We denote $\mathcal{T}_i$ the time steps at which robot i propagates its latest map to the network, i.e., if robot i decides it wants to initiate the propagation of its latest map, then $k + 1 \in \mathcal{T}_i$; otherwise, $k + 1 \notin \mathcal{T}_i$ and robot i keeps on merging the previous map. We let d_i be the degree of a robot i, containing the total number of times its local map changes (the cardinality of $\mathcal{T}_i$), and d be the degree of the team,

$$d_i = |\mathcal{T}_i|, \qquad\qquad d = d_1 + \cdots + d_n. \qquad (4.6)$$

In this paper we consider that the number of times robots propagate the changes of their local maps d is finite. These changes give rise to a different global map (Eq. (4.5)) and robots must update their estimates to react to this change.

Problem 1 We consider $n \in \mathbb{N}$ robots exploring and acquiring local maps at some time steps k as in Eqs. (4.1), (4.2). The communication is range-limited and two robots can exchange data only if they are close enough. We let $\mathcal{G}_k = (\mathcal{V}, \mathcal{E}_k)$ be the undirected communication graph at step k. The nodes are the robots, $\mathcal{V} = \{1, \ldots, n\}$. If robots i, j can communicate then there is an edge between them, $(i, j) \in \mathcal{E}_k$. The set of neighbors $\mathcal{N}_i^k$ of robot i at step k is

$$\mathcal{N}_i^k = \{j \mid (i, j) \in \mathcal{E}_k, j \neq i\}.$$

The goal is the design of distributed algorithms so that each robot $i \in \mathcal{V}$ computes and tracks the global map in Eq. (4.5), and the blocks in the main diagonal of its covariance matrix, based on local interactions with its neighbors $\mathcal{N}_i^k$. ☐

4.3 Dynamic Map Merging Algorithm

The space-time diffusion methods have been previously used under independent observations of static variables [51]. In our map merging scenario, the map features $\mathbf{x}$ are static but the robot poses $\mathbf{r}_i^k$ vary with time k. Besides, the local map of a robot i at step k is an evolution of its local map at previous steps $k' < k$. Thus, the local maps $Y_i^k, \mathbf{y}_i^k$ (Eq. (4.4)) are not independent and this has to be taken into account, because otherwise the same information would be considered several times. For the previous reasons, we propose to use space-time diffusion ideas using as inputs the *information increments* associated to the feature estimates instead of the maps $Y_i^k, \mathbf{y}_i^k$.

We first pay attention to Eq. (4.5). Using classical matrix block-wise inversion rules [21, Chap. 0.7], the global estimates $\hat{\mathbf{x}}_G^k$ of the positions of the static features $\mathbf{x}$ in Eq. (4.5), and its associated block $\Sigma_G^k \doteq E[\hat{\mathbf{x}}_G^k(\hat{\mathbf{x}}_G^k)^T]$ within the covariance matrix $(\sum_{i=1}^n Y_i^k)^{-1}$ are given by

$$\hat{\mathbf{x}}_G^k = \left(I_G^k\right)^{-1} \mathbf{i}_G^k, \qquad\qquad \Sigma_G^k = \left(I_G^k\right)^{-1}, \qquad (4.7)$$

where $I_G^k \in \mathbb{R}^{\mathcal{M} \times \mathcal{M}}, \mathbf{i}_G^k \in \mathbb{R}^{\mathcal{M}}$ are the information matrix and vector of the estimates of the features' positions $\mathbf{x}$ in the global map at step k in IF form,

$$I_G^k = \sum_{i=1}^n I_i^k, \qquad\qquad \mathbf{i}_G^k = \sum_{i=1}^n \mathbf{i}_i^k, \qquad (4.8)$$

and $I_i^k \in \mathbb{R}^{\mathcal{M} \times \mathcal{M}}$ and $\mathbf{i}_i^k \in \mathbb{R}^{\mathcal{M}}$ are the information matrix and vector of the local estimates $\hat{\mathbf{x}}_i^k$ of the features' positions $\mathbf{x}$ in the local map (Eq. (4.1)) at robot $i \in \{1, \ldots, n\}$ and step k in IF form,

$$I_i^k = \left(H_i^k\right)^T \left(\Sigma_i^k\right)^{-1} H_i^k, \qquad\qquad \mathbf{i}_i^k = \left(H_i^k\right)^T \left(\Sigma_i^k\right)^{-1} \hat{\mathbf{x}}_i^k. \qquad (4.9)$$

The global estimates $\hat{\mathbf{r}}_{G,i}^k$ of the set of poses $\mathbf{r}_i^k$ of each robot i up to step k in Eq. (4.5), and its associated block $R_{G,ii}^k \doteq E[\hat{\mathbf{r}}_{G,i}^k(\hat{\mathbf{r}}_{G,i}^k)^T]$ within the covariance matrix $(\sum_{i=1}^n Y_i^k)^{-1}$, can be obtained from $\hat{\mathbf{x}}_G^k$, Σ_G^k (Eq. (4.7)) and from the local maps $\hat{\mathbf{r}}_i^k$, $\hat{\mathbf{x}}_i^k$, R_i^k, S_i^k, Σ_i^k, H_i^k (Eqs. (4.1), (4.2)) as follows:

$$\hat{\mathbf{r}}_{G,i}^k = \hat{\mathbf{r}}_i^k + S_i^k \left(\Sigma_i^k\right)^{-1} \left(H_i^k \hat{\mathbf{x}}_G^k - \hat{\mathbf{x}}_i^k\right),$$
$$R_{G,ii}^k = R_i^k - S_i^k \left(\Sigma_i^k\right)^{-1} \left(S_i^k\right)^T$$
$$+ S_i^k \left(\Sigma_i^k\right)^{-1} H_i^k \Sigma_G^k \left(H_i^k\right)^T \left(\Sigma_i^k\right)^{-1} \left(S_i^k\right)^T. \qquad (4.10)$$

Here we are interested (Problem 1) in computing the blocks in the main diagonal of the covariance matrix $(\sum_{i=1}^n Y_i^k)^{-1}$. The expressions for the off-diagonal terms can be found, e.g., in [4].

Thus, the original problem can be decomposed into two parts: the estimation of the features' positions (Eqs. (4.7)–(4.9)), which requires the robots to reach consensus on the information matrices and vectors of the features' positions (Eq. (4.8)); and the estimation of the robot poses (Eq. (4.10)), that only requires information local to each robot i, and the features estimates $\hat{\mathbf{x}}_G^k$, Σ_G^k. We propose an algorithm that consists of keeping up-to-date estimates of the features' positions $\hat{\mathbf{x}}_G^k$, Σ_G^k, using dynamic average consensus on the information increments of the local information matrices associated to features' positions I_i^k, $\mathbf{i}_i^k$ in Eq. (4.9). At each step k, each robot i uses its most recent estimates of $\hat{\mathbf{x}}_G^k$ and Σ_G^k to obtain the estimates of its robot poses $\hat{\mathbf{r}}_{G,i}^k$, $R_{G,ii}^k$ (Eq. (4.10)) and propagates this vector $\hat{\mathbf{r}}_{G,i}^k$ and the main diagonal elements of matrix $R_{G,ii}^k$ to the remaining robots in the network. In the remaining of this section, we deeply discuss the part concerning the consensus on the information increments. We revise properties of convergence and unbiased mean. We carry out a careful study to show that the estimates are consistent.

For each robot $i \in \{1, \ldots, n\}$ we define the following *increment* information matrix $\Delta_i^k \in \mathbb{R}^{\mathcal{M} \times \mathcal{M}}$ and vector $\delta_i^k \in \mathbb{R}^{\mathcal{M}}$,

$$\begin{aligned}
\Delta_i^k &= I_i^k - I_i^{k-1}, & \delta_i^k &= \mathbf{i}_i^k - \mathbf{i}_i^{k-1}, & \text{for } k \geq 1, \\
\Delta_i^k &= I_i^k, & \delta_i^k &= \mathbf{i}_i^k, & \text{for } k = 0.
\end{aligned} \tag{4.11}$$

Note that for all the robots such that $k \notin \mathcal{T}_i$, the increment information matrices and vectors will be zero. The associated features' position estimates within the global map at step k in Eq. (4.8) can be expressed in terms of the previous global estimate at step $k-1$ and the map increments at k as follows:

$$I_G^k = I_G^{k-1} + \sum_{i=1}^n \Delta_i^k, \qquad \mathbf{i}_G^k = \mathbf{i}_G^{k-1} + \sum_{i=1}^n \delta_i^k. \tag{4.12}$$

Equivalently, the estimates of the features' positions in the local map of each robot i at step k, and in the global map at step k can be expressed in terms of the map increments at all the previous steps $k' = 0, \ldots, k$,

$$\begin{aligned}
I_i^k &= \sum_{k'=0}^k \Delta_i^{k'}, & \mathbf{i}_i^k &= \sum_{k'=0}^k \delta_i^{k'}, \\
I_G^k &= \sum_{i=1}^n \sum_{k'=0}^k \Delta_i^{k'} & \mathbf{i}_G^k &= \sum_{i=1}^n \sum_{k'=0}^k \delta_i^{k'}.
\end{aligned} \tag{4.13}$$

Each robot i maintains an estimate of the averaged information matrix $\hat{I}_i^A(k) \in \mathbb{R}^{\mathcal{M} \times \mathcal{M}}$ and vector $\hat{\mathbf{i}}_i^A(k) \in \mathbb{R}^{\mathcal{M}}$, and of its degree $d_i(k)$ containing the number of times it has updated its local map; recall that each robot i propagates the changes in its local map at specific and locally decided time steps $k \in \mathcal{T}_i$. Robot $i \in \{1, \ldots, n\}$ initializes its variables with

$$d_i(-1) = 0, \qquad \hat{I}_i^A(0) = \mathbf{0}, \qquad \hat{\mathbf{i}}_i^A(0) = \mathbf{0}, \tag{4.14}$$

and updates them at all $k \geq 0$ with the following algorithm [5, 7].

Algorithm 5 Dynamic map merging—robot i, iteration k.
(Measurement update:)

$$\text{If } k \in \mathcal{T}_i, d_i(k) = d_i(k-1) + 1,$$

$$\hat{I}_i^A(k_+) = (1 - 1/d_i(k))\hat{I}_i^A(k) + \Delta_i^k/d_i(k),$$

$$\hat{\mathbf{i}}_i^A(k_+) = (1 - 1/d_i(k))\hat{\mathbf{i}}_i^A(k) + \delta_i^k/d_i(k); \tag{4.15}$$

$$\text{otherwise, } d_i(k) = d_i(k-1),$$

$$\hat{I}_i^A(k_+) = \hat{I}_i^A(k), \hat{\mathbf{i}}_i^A(k_+) = \hat{\mathbf{i}}_i^A(k). \tag{4.16}$$

(Spatial update:) If $d_i(k) > 0$,

$$\hat{I}_i^A(k+1) = \sum_{j \in \mathcal{N}_i^k \cup \{i\}} \mathcal{W}_{ij}(k)\hat{I}_j^A(k_+),$$

$$\hat{\mathbf{i}}_i^A(k+1) = \sum_{j \in \mathcal{N}_i^k \cup \{i\}} \mathcal{W}_{ij}(k)\hat{\mathbf{i}}_j^A(k_+), \tag{4.17}$$

where the space-time weight matrix $\mathcal{W}(k) \in \mathbb{R}^{n \times n}$ is

$$\mathcal{W}_{ii}(k) = 1 - \sum_{j \in \mathcal{N}_i^k} \mathcal{W}_{ij}(k), \quad \text{and for } j \neq i$$

$$\mathcal{W}_{ij}(k) = d_j(k)/\max\{d_i^{st}(k), d_j^{st}(k)\}, \text{ if } (i, j) \in \mathcal{E}_k,$$

$$\mathcal{W}_{ij}(k) = 0 \quad \text{otherwise}, \tag{4.18}$$

and where $d_i^{st}(k)$ is the space-time degree of each robot i at step k, containing the number of map changes propagated by both robot i and its neighbors $\mathcal{N}_i^k$ up to step k, $d_i^{st}(k) = \sum_{j \in \mathcal{N}_i^k \cup \{i\}} d_j(k)$. □

Robots decide on their own when they want to execute a new measurement update step. If up to step k a robot i never tried to merge its map with the other robots, then $d_i(k) = 0$, and thus it does not execute the spatial update. We consider these robots i as disconnected from the others ($(i, j) \notin \mathcal{E}_k$ for all $j \neq i$), even if they are in communication range.

The superscript A in the variables of the previous algorithm refers to the fact that the matrices and vectors estimated by the robots track the *average* of the information increments, instead of its sum. In several places in this paper, we will refer instead to the *global* estimates of a robot i, which are obtained from the averaged variables $\hat{I}_i^A(k), \hat{\mathbf{i}}_i^A(k)$ as follows,

$$\hat{I}_i^G(k) = d(k)\hat{I}_i^A(k), \qquad \hat{\mathbf{x}}_i^G(k) = \left(\hat{I}_i^A(k)\right)^{-1}\hat{\mathbf{i}}_i^A(k),$$

$$\hat{\mathbf{i}}_i^G(k) = d(k)\hat{\mathbf{i}}_i^A(k), \qquad \hat{\Sigma}_i^G(k) = \left(\hat{I}_i^A(k)\right)^{-1}/d(k), \tag{4.19}$$

where $d(k) = d_1(k) + \cdots + d_n(k)$ is the degree of the robot team containing the number of times robots propagated changes in their local maps, up to step k. Note that in order to obtain the global map estimate in Information Filter form $\hat{I}_i^G(k)$, $\hat{\mathbf{i}}_i^G(k)$, or to compute the covariance matrix $\hat{\Sigma}_i^G(k)$, robots need to estimate in parallel the total amount of measurement update steps $d(k)$. This can be done, for instance, using similar techniques as for estimating the number of nodes [28, 47]. Later in this section (Theorem 8) we provide an expression $(\hat{I}_i^A(k+1))^{-1}/d_i(k)$ for the covariance that ensures it remains consistent. When robots compute the unbiased mean $\hat{\mathbf{x}}_i^G(k)$ (Eq. (4.19)) and consistent covariance $(\hat{I}_i^A(k+1))^{-1}/d_i(k)$ (Eq. (4.26)) form, they do not need to know $d(k)$, but only $d_i(k)$ which is local to each robot.

4.4 Properties of the Dynamic Map Merging Algorithm

An interesting property of this map merging algorithm is that the temporal global maps $\hat{\mathbf{x}}_i^k(t)$ estimated at each robot i, are unbiased estimates of the true feature positions $\mathbf{x}$ and the covariance matrices are consistent [5, 7]. As a result, the robots do not need to wait for any specific number of iterations of the map merging algorithm. Instead, they can make decisions on their temporal global map estimates whenever they need.

Lemma 2 (Convergence) *Assume all the robots $i \in \mathcal{V}$ execute the dynamic map merging algorithm (Algorithm 5), and assume that the set of communication graphs that occur infinitely often is jointly connected. Let $k_\star \geq \max\{k \in \mathcal{T}_i\}$ for all $i \in \mathcal{V}$ be a step when all map updates have been propagated by the robots, and $\hat{\mathbf{x}}_G^{k_\star}$, $\Sigma_G^{k_\star}$ be the centralized global estimate of the features' positions at this step, ($I_G^{k_\star}$, $\mathbf{i}_G^{k_\star}$ in IF form), given by Eqs. (4.5), (4.7)–(4.9). Then, the estimated information matrix $\hat{I}_i^G(k)$, information vector $\hat{\mathbf{i}}_i^G(k)$, mean $\hat{\mathbf{x}}_i^G(k)$, and covariance $\hat{\Sigma}_i^G(k)$ as in Eq. (4.19), at each robot $i \in \mathcal{V}$ asymptotically converge to this global estimate,*

$$\lim_{k\to\infty} \hat{I}_i^G(k) = I_G^{k_\star}, \qquad\qquad \lim_{k\to\infty} \hat{\mathbf{x}}_i^G(k) = \hat{\mathbf{x}}_G^{k_\star},$$

$$\lim_{k\to\infty} \hat{\mathbf{i}}_i^G(k) = \mathbf{i}_G^{k_\star}, \qquad\qquad \lim_{t\to\infty} \hat{\Sigma}_i^G(k) = \Sigma_G^{k_\star}. \qquad (4.20)$$

Proof As it is stated by [51, Th. 2], if the set of communication graphs $\mathcal{G}_k$ that occur infinitely often is jointly connected, then

$$\lim_{k\to\infty} \hat{I}_i^A(k) = \sum_{i=1}^n \sum_{k'\in\mathcal{T}_i} \frac{\Delta_i^{k'}}{d}, \qquad\qquad \lim_{k\to\infty} \hat{\mathbf{i}}_i^A(k) = \sum_{i=1}^n \sum_{k'\in\mathcal{T}_i} \frac{\delta_i^{k'}}{d},$$

which equals $I_G^k/d(k)$, $\mathbf{i}_G^k/d(k)$ in Eq. (4.8) when all measurements have been taken, i.e., when $k_\star \geq \max\{k \in \mathcal{T}_i\}$ for all $i \in \mathcal{V}$. $\qquad\qquad\square$

Now, we present a more compact expression for Eqs. (4.15)–(4.17) in Algorithm 5 which will simplify the analysis of the remaining properties:

$$d_i(k)\hat{I}_i^A(k+1) = \sum_{j=1}^n \mathscr{W}(k)_{ji} d_j(k-1)\hat{I}_j^A(k)$$

$$+ \sum_{j=1}^n \mathscr{W}(k)_{ji}(d_j(k) - d_j(k-1))\Delta_j^k, \qquad (4.21)$$

Moreover, since $\hat{I}_i^A(0) = \mathbf{0}$, then

$$d_i(k)\hat{I}_i^A(k+1) = \sum_{k'=0}^k \sum_{j=1}^n [\Phi(k,k')]_{ij} \Delta_j^{k'}, \qquad (4.22)$$

where matrix $\Phi(k,k')$, with $k' \leq k$, is

$$\Phi(k,k') = \mathscr{W}(k)^T \dots \mathscr{W}(k')^T (D(k') - D(k'-1)), \qquad (4.23)$$

and $D(k) \in \mathbb{R}^{n \times n}$ is a diagonal matrix; each entry of its main diagonal $D_{ii}(k)$ equals the degree $d_i(k)$ of robot i at step k. The equivalent expressions for $\hat{\mathbf{i}}_i^A(k+1)$ are got by replacing $\hat{I}_j^A(k)$, Δ_j^k with $\hat{\mathbf{i}}_j^A(k)$, δ_j^k.

Lemma 3 (Unbiased mean) *The estimates of the features' positions in the global map mean* $\hat{\mathbf{x}}_i^G(k)$, *for each robot* $i \in \mathcal{V}$, *after* k *iterations of Algorithm 5, such that* $d_i(k-1) > 0$, *are unbiased estimates of the true feature positions* $\mathbf{x}$,

$$\mathrm{E}\left[\hat{\mathbf{x}}_i^G(k)\right] = \mathrm{E}\left[\left(\hat{I}_i^A(k)\right)^{-1}\hat{\mathbf{i}}_i^A(k)\right] = \mathbf{x}. \qquad (4.24)$$

Proof It can be done in a similar fashion as in [51] by noting that the local features' positions estimates $\hat{\mathbf{x}}_j^k$ at each robot j (Eq. (4.1)) are an observation of the true $\mathbf{x}$,

$$\hat{\mathbf{x}}_j^k = H_i^k \mathbf{x}_G + \mathbf{v}_j^k, \text{ with} \qquad \mathrm{E}\left[\mathbf{v}_j^k\right] = 0,$$

and the increment information vector $\delta_j^k = \mathbf{i}_j^k - \mathbf{i}_j^{k-1}$ is

$$\delta_j^k = \left(H_j^k\right)^T \left(\Sigma_j^k\right)^{-1} \mathbf{v}_j^k - \left(H_j^{k-1}\right)^T \left(\Sigma_j^{k-1}\right)^{-1} \mathbf{v}_j^{k-1} + \Delta_j^k \mathbf{x},$$

which combined with Eq. (4.22) gives

$$\hat{\mathbf{i}}_i^A(k) = \hat{I}_i^A(k)\mathbf{x} + \frac{1}{d_i(k-1)} \sum_{k'=1}^{k-1} \sum_{j=1}^{n} [\Phi(k-1,k')]_{ij}$$

$$\left(\left(H_j^{k'}\right)^T \left(\Sigma_j^{k'}\right)^{-1} \mathbf{v}_j^{k'} - \left(H_j^{k'-1}\right)^T \left(\Sigma_j^{k'-1}\right)^{-1} \mathbf{v}_j^{k'-1} \right),$$

and thus $E[(\hat{I}_i^A(k))^{-1}\hat{\mathbf{i}}_i^A(k)] = \mathbf{x}$ since the noises $\mathbf{v}_j^{k'}$ have zero mean for all k and all $j \in \mathscr{V}$. □

Next we present our main result, regarding the consistency of the maps estimated by the robots, at each iteration. This property is of high interest in map merging scenarios. This means that at each step k, robots have indeed a map that they can use. As a result, the robots do not need to wait for any specific number of iterations of the map merging algorithm. Instead, they can make decisions on their temporal global map estimates whenever they need. Our result relies on condition $I_j^{k+1} \succeq I_j^k$, which means that the local estimates of the features' positions at successive steps have more information, or equivalently, that they become more precise. Note that this is the behavior expected in classical SLAM approaches [17] as more observations are taken, and in our experiments it has been always observed. There is an additional condition, $d_i(k) > 0$; recall that $d_i(k) = 0$ means that robot i has not initiated the map merging process yet. Since robot i has not computed any covariance yet, it does not make sense to question whether its covariance is consistent or not.

Theorem 8 (Consistent covariance) *Assume that the local map at each robot j satisfies, for successive steps k, $k + 1$,*

$$I_j^{k+1} \succeq I_j^k, \tag{4.25}$$

Then, the covariance $(\hat{I}_i^A(k+1))^{-1}/d_i(k)$ estimated by each robot i for which $d_i(k) > 0$, at each iteration k, is consistent with respect to the centralized covariance matrix Σ_G^k,

$$\left(\hat{I}_i^A(k+1)\right)^{-1}/d_i(k) \succeq \Sigma_G^k. \tag{4.26}$$

Proof Along this proof we use the following change of variables; we let $\hat{J}_i^A(k)$ be

$$\hat{J}_i^A(k) = d_i(k-1)\hat{I}_i^A(k), \tag{4.27}$$

and note that if $d_i(k-1) = 0$, then $\hat{J}_i^A(k) = \mathbf{0}$. From Eq. (4.21), this variable evolves according to

$$\hat{J}_i^A(k+1) = \sum_{j=1}^{n} \mathcal{W}_{ji}(k)\hat{J}_j^A(k)$$

$$+ \sum_{j=1}^{n} \mathcal{W}_{ji}(k)(d_j(k) - d_j(k-1)) \left(I_j^k - I_j^{k-1} \right), \qquad (4.28)$$

where $d_j(k) - d_j(k-1) = 1$ if robot j introduced a new map increment during the last step k, and zero otherwise, and $\mathcal{W}(k)_{ji}$, is given by Eq. (4.18). Note that the entries of matrix $\mathcal{W}(k)$ are numbers between 0 and 1, and recall that $\mathcal{W}_{ji}(k) = 0$ if $d_i(k) = 0$ or $d_j(k) = 0$.

We want to prove that, for all i and k,

$$\hat{J}_i^A(k+1) \preceq \sum_{j=1}^{n} I_j^k = I_G^k; \qquad (4.29)$$

this is done by induction. We consider first that case $k = 0$, where the robots states $\hat{J}_j^A(k)$ are initialized with zeros, where $d_j(-1) = 0$, and where the map increments are exactly the maps at $k = 0$, since $I_j^{k-1} = \mathbf{0}$ for $k = 0$; we have that for all i,

$$\hat{J}_i^A(1) = \sum_{j=1}^{n} \mathcal{W}_{ji}(0)d_j(0)I_j^0. \qquad (4.30)$$

Since the weights $\mathcal{W}_{ji}(k)$ are numbers between 0 and 1, the degrees $d_j(0)$ are equal to 0 or to 1, and the local information matrices I_j^k are positive semidefinite, $I_j^k \succeq 0$, then we have $\mathcal{W}_{ji}(k)d_j(k)I_j^k \preceq I_j^k$, and thus

$$\hat{J}_i^A(1) \preceq \sum_{j=1}^{n} I_j^0 = I_G^0. \qquad (4.31)$$

Now that we have proved that it is true for $k = 0$, we assume it is true for k, i.e., $\hat{J}_j^A(k) \preceq I_G^{k-1} = \sum_{j'=1}^{n} I_{j'}^{k-1}$ for all j, and we try to prove than then it holds for $k + 1$ as well. Considering Eq. (4.28), and taking into account that the weights satisfy $\sum_{j=1}^{n} \mathcal{W}_{ji}(k) = 1$, we have

$$\hat{J}_i^A(k+1) = \sum_{j=1}^{n} \mathcal{W}_{ji}(k)\hat{J}_j^A(k)$$

$$+ \sum_{j=1}^{n} \mathcal{W}_{ji}(k)(d_j(k) - d_j(k-1)) \left(I_j^k - I_j^{k-1} \right)$$

$$\leq \sum_{j=1}^{n} \mathscr{W}_{ji}(k) \left(\sum_{j'=1}^{n} I_{j'}^{k-1} \right) + \sum_{j=1}^{n} \mathscr{W}_{ji}(k) \left(I_j^k - I_j^{k-1} \right)$$

$$= \sum_{j'=1}^{n} I_{j'}^{k-1} + \sum_{j=1}^{n} \mathscr{W}_{ji}(k) \left(I_j^k - I_j^{k-1} \right). \tag{4.32}$$

From condition (4.25), $I_i^k - I_i^{k-1} \succeq \mathbf{0}$, and thus, using again the fact that $\mathscr{W}_{ji}(k)$ are positive numbers between 0 and 1, we have

$$\hat{J}_i^A(k+1) \preceq \sum_{j=1}^{n} I_j^{k-1} + \sum_{j=1}^{n} \left(I_j^k - I_j^{k-1} \right) = \sum_{j=1}^{n} I_j^k = I_G^k, \tag{4.33}$$

concluding that $\hat{J}_i^A(k+1) \preceq I_G^k$. Thus, when $d_i(k) > 0$,

$$\frac{\left(\hat{I}_i^A(k+1) \right)^{-1}}{d_i(k)} = \left(\hat{J}_i^A(k+1) \right)^{-1} \succeq \left(I_G^k \right)^{-1} = \Sigma_G^k, \tag{4.34}$$

which concludes the proof. $\qquad\square$

Note that the results about the estimated merged maps being unbiased and consistent (Lemma 3 and Theorem 8) rely on the local maps being consistent as in Eqs. (4.1) and (4.2). Depending on the sensing model, e.g., if robots can only obtain partial observations of the features positions, and depending on the local mapping method used, the local maps may not be consistent. Even in this case, the global maps estimated by our algorithm are more conservative than the centralized map.

We finally note that the estimates of the robot poses $\hat{\mathbf{r}}_{G,i}^k$, $R_{G,ii}^k$ are obtained by each robot i by replacing $\hat{\mathbf{x}}_G^k$ and Σ_G^k in Eq. (4.10) with its most recent estimates of the features' positions. It can be easily checked that by using $\hat{\mathbf{x}}_i^G(k)$, $\hat{\Sigma}_i^G(k)$ (Eq. (4.19)), the estimates of $\hat{\mathbf{r}}_{G,i}^k$, $R_{G,ii}^k$ are convergent as in Lemma 2; and by using $\hat{\mathbf{x}}_i^G(k)$ and the expression for the consistent covariance $(\hat{I}_i^A(k+1))^{-1}/d_i(k)$ the estimates of $\hat{\mathbf{r}}_{G,i}^k$, $R_{G,ii}^k$ are unbiased and consistent, as in Lemma 3 and Theorem 8.

Communication and Memory Costs

Now we discuss what are the benefits of using consensus-based approaches instead of classical propagation methods, in terms of communication and memory costs.

Several distributed map merging methods rely on propagating local data whenever this data changes, e.g., raw data or local map representations. The ones based on raw (not processed) data, have several inconveniences, and they usually present large memory and communication costs. The ones that propagate local maps seem appealing from the communication point of view, since each piece of data traverses the network only once, whereas consensus-based methods transmit information at each iteration. However, methods based on propagating local maps have the inconvenience that, in addition to the global map, each robot must store the local map of every other robot in the network. Note that we are considering scenarios where the communication network can get disconnected at any moment, and individual or small groups of robots can leave the remaining team for long periods of time. In order to properly re-synchronize with them in posterior meetings, and correctly replace the old information in the global map, robots must keep track of all the information (local maps) available. Thus, the memory cost is $\sum_{i=1}^{n}((\mathscr{R}_i^k + \mathscr{M})^2 + (\mathscr{R}_i^k + \mathscr{M}))$ for storing either the n information matrices and vectors, or the n mean vectors and covariance matrices,[3] plus $(\mathscr{M}_G^k)^2 + \mathscr{M}_G^k$ for the global map. The memory cost does not scale well with the size of the network, i.e., if the number of robots is increased without changing the scene size, the memory cost increases as well. Consensus-based approaches do not suffer from this problem, since each robot keeps a single representation of the scene, and thus the memory cost does not depend on the number of nodes.

Similarly to the memory cost discussion, in consensus-based approaches, robots send their single representation of the scene at each iteration, so that the communication cost per iteration exclusively depends on the size of the scene, and it is almost equal for all the robots. However, propagation methods do not have any control about the amount of new information that arrives to a particular robot; thus, they are prone to generate high communications peaks and bottlenecks in some areas of the network. The communication load is not properly balanced, so some particular robots may be sending large amounts of data. Due to the iterative nature of consensus methods, the total final communication cost may be larger than for other approaches depending on the number of iterations executed by the robots. This convergence speed depends on the network topology and it is related to the algebraic connectivity of the communication graph, as discussed later in this section.

Thus, using consensus strategies is a more efficient choice whenever there is common information that was observed by several robots, whereas propagation methods make sense when there is no overlapping in the features observed by the robot team. Our method combines the benefits of both approaches: consensus is executed to estimate the feature positions $\hat{\mathbf{x}}_G^k$, Σ_G^k, with memory cost $(\mathscr{M})^2 + \mathscr{M}$ and communication cost per iteration $(\mathscr{M})^2 + \mathscr{M}$; and each robot i locally estimates its poses $\hat{\mathbf{r}}_{G,i}^k$, $R_{G,ii}^k$ (Eq. (4.10)) and propagates vector $\hat{\mathbf{r}}_{G,i}^k$ and the main diagonal of $R_{G,ii}^k$. Thus, the memory cost per robot for storing the global map is $(\sum_{i=1}^{n} \mathscr{R}_i^k) + (\mathscr{M})^2 + \mathscr{M}_G^k$, and there is no need to keep any additional information from the other robots. The communication cost associated to the propagation (vector $\hat{\mathbf{r}}_{G,i}^k$ and the main diagonal of $R_{G,ii}^k$) is light, since these elements are vectors. Moreover, in practice, our robots execute the algorithm described in this paper for estimating the global mean $\hat{\mathbf{x}}_G^k$ and

---

[3]$(\mathscr{R}_i^k + \mathscr{M})^2$ is a worst case cost for the information matrices; in practical applications, a better performance can be achieved by taking advantage of their sparse structure. E.g., for full robot trajectories approaches, it can be order $(\mathscr{M} + (l + 1)\mathscr{R}_i^k)$, where l is the average number of features observed from each robot pose.

covariance Σ_G^k of the *common* features, i.e., using the information increments of the features that appear in several local maps. In addition, a robot i may have been the only one that has observed some exclusive features. These exclusive features are managed in the same fashion as for the estimated robot poses, i.e., they are reestimated and its mean and the main diagonal entries of their covariance matrix are propagated. As a result, all the robots have the information of the exclusive features of the other robots. Thus, the size $\mathcal{M}$ used in this paragraph refers to the number of common features, and the sizes $\mathcal{R}_i^k$ to the number of poses and exclusive features at robot i. Equivalently, the computational cost of our method, which is cubic on the size $\mathcal{M}$ (Eq. (4.19)), refers to the number of common features as well.

Initial Correspondence and Data Association

The expressions in Eqs. (4.4), (4.7)–(4.10) implicitly assume that the local maps are expressed in a common reference frame. This issue is related to initial correspondence, network localization, or map alignment problems. A discussion of different methods for computing this network localization can be found in Chap. 3.

Equivalently, for simplicity, we have presented the formulation in Sects. 4.2 and 4.3 including the structures of the information matrices and vectors $\hat{\mathbf{i}}_i^A(k)$, $\hat{I}_i^A(k)$, as if robots knew the total amount of features m and the relationship between their local features and the global ones, encoded in the observation matrices H_i^k in Eq. (4.9). The problem of establishing a relationship between the elements observed by the different robots is known as data association, and it is discussed in Chap. 2. First, local matches are established between the variables of neighboring sensors; after that, exclusive variables are identified without requiring any extra efforts: they are variables that have not been associated to any other one. Robots discover the features observed by the others in the messages exchanged at each iteration, and introduce new columns and rows in $\hat{\mathbf{i}}_i^A(k)$, $\hat{I}_i^A(k)$ accordingly. As a result, the information matrices and vectors do not contain non-informative zero rows and columns. Information matrices $\hat{I}_i^A(k)$ (Eq. (4.19)) are invertible at each iteration of the algorithm and thus the global map can always be estimated. Note also that the total number of features m is used only as a tool for presenting the formulation, but it does not need to be known by the robots or even to be fixed. Instead, the variables managed by the robots $\hat{\mathbf{i}}_i^A(k)$, $\hat{I}_i^A(k)$ have a structure that is adapted according to the features observed by the robot team.

Convergence Speed

As we mentioned before, the consensus is asymptotically reached, which means that the time until completion is infinite. The convergence speed depends on the network topology and it is related to the algebraic connectivity of the communication graph, as discussed in [2, 3]. There exist several methods for estimating this algebraic connectivity, e.g., [6]. The number of iterations can also be easily optimized in a local way, by executing a new consensus iteration only if the neighborhood has changed, or if there have been great modifications in the state of some of the robots in the neighborhood. Here, we provide some well known convergence speed results for the case that the local maps do not change along the merging. These results only depend on the graph topology. For dynamic map merging scenarios, the characterization of this speed strongly depends on the changes in the local maps as well.

The convergence speed of the averaging algorithm presents a geometric rate for fixed graphs [10, 50] which depends on the second eigenvalue with the largest absolute value $|\lambda_2(W)|$ in the Metropolis weights matrix (Eq. (A.3) in Appendix A). If we denote $\gamma = |\lambda_2(W)|$, it can be shown that each entry $[\hat{I}_i^G(k)]_{r,s}$, $[\hat{\mathbf{i}}_i^G(k)]_r$ in the information matrices and vectors estimated by the robots evolve according to

$$|[\hat{I}_i^G(k)]_{r,s} - [I_G^k]_{r,s}| \le (\gamma)^k \sqrt{n} \max_j \left\{ \left| [\hat{I}_j^G(0)]_{r,s} - [I_G^k]_{r,s} \right| \right\}, \qquad (4.35)$$

$$|[\hat{\mathbf{i}}_i^G(k)]_r - [\mathbf{i}_G^k]_r| \le (\gamma)^t \sqrt{n} \max_j \left\{ \left| [\hat{\mathbf{i}}_j^G(0)]_r - [\mathbf{i}_G^k]_r \right| \right\}, \qquad (4.36)$$

for all $i \in \{1, \dots, n\}$, all $r, s \in \{1, \dots, \mathcal{M}\}$, and all $k \ge 0$.

For graphs with switching topology $\mathscr{G}_k = (\mathcal{V}, \mathscr{E}_k)$, the convergence speed is geometric if the graph has an interval of joint connectivity τ such that every subsequence

$$\{G_{t_0+1}, \dots, G_{t_0+\tau}\}$$

of length τ is jointly connected for all t_0 [10]. In these graphs, the τ-index of joint contractivity $\delta < 1$ is given by

$$\delta = \max_{\mathscr{W} \in \mathbb{W}_\tau} \{|\lambda_2(\mathscr{W})| | \mathscr{W} \text{ primitive paracontractive}\}, \qquad (4.37)$$

where $\mathbb{W}_\tau$ is the set of all products of, at most, τ Metropolis matrices $W(t)$ that can be obtained in the communication graph. The convergence speed of each entry $[\hat{I}_i^G(k)]_{r,s}$, $[\hat{\mathbf{i}}_i^G(k)]_r$ in the information matrices and vectors estimated by the robots depends on the τ-index of joint contractivity,

$$|[\hat{I}_i^G(k)]_{r,s} - [I_G^k]_{r,s}| \le (\delta)^{\lfloor \frac{k}{\tau} \rfloor} \sqrt{n} \max_j \left\{ \left| [\hat{I}_j^G(0)]_{r,s} - [I_G^k]_{r,s} \right| \right\}, \qquad (4.38)$$

$$|[\hat{\mathbf{i}}_i^G(k)]_r - [\mathbf{i}_G^k]_r| \le (\delta)^{\lfloor \frac{k}{\tau} \rfloor} \sqrt{n} \max_j \left\{ \left| [\hat{\mathbf{i}}_j^G(0)]_r - [\mathbf{i}_G^k]_r \right| \right\}, \qquad (4.39)$$

where $\lfloor \frac{k}{\tau} \rfloor$ is the largest integer less than or equal to $\frac{k}{\tau}$.

Therefore, the convergence speed depends on the topology of the communication graph. Moreover, from the time complexity analysis, we can see that when the communication graph is complete, the robots reach consensus in one iteration. For complete communication graphs, the Metropolis matrix is $W = (1/n)\mathbf{1}\mathbf{1}^T$ and $|\lambda_2(W)| = 0$. Then, $|[\hat{I}_i^G(k)]_{r,s} - [I_G^k]_{r,s}| \le 0$, $|[\hat{\mathbf{i}}_i^G(k)]_r - [\mathbf{i}_G^k]_r| \le 0$ for all $k \ge 1$, all $i \in \{1, \dots, n\}$, and all $r, s \in \{1, \dots, \mathcal{M}\}$.

4.5 Simulations

We have performed Monte Carlo simulations with 5 robots following the trajectories in Fig. 4.1a. They start in the right part of the scenario and finish in the left part. We consider a $10 \times 10 \times 10$ m scenario with features spread over two walls and the floor. Three of the robots observe the walls and two of them the floor at different

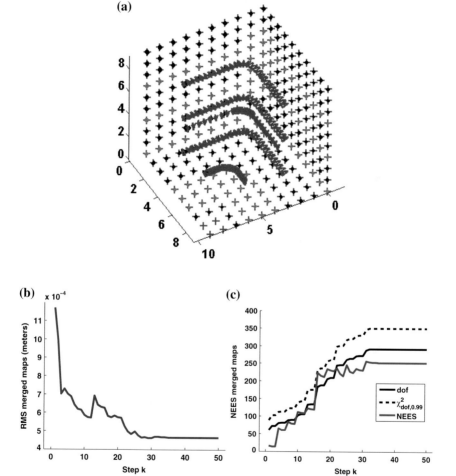

Fig. 4.1 **a** Scenario. 5 robots observing features; 3 robots point toward the walls, and 2 observe the floor. *Red crosses* are the ground-truth position of the features. *Red triangles* represent the ground-truth robot trajectories. *Gray* and *black dots* are the estimated position of the features observed respectively by several robots and by only one robot. For each feature, we display the 100 points obtained during the 100 Monte Carlo simulations. **b** Average root mean square error (RMS), Eq. (4.40). **c** Average normalized state estimation error squared (NEES), Eq. (4.40)

heights. Red crosses represent the ground-truth position of the features, and red triangles the ground-truth robot trajectories. Robots measure features that have a depth between 0.4 and 5 m, and which are placed in front of them and within the image limits. These observations are corrupted with noises with standard deviation $0.0012 + 0.0019(depth - 0.4)^2$ for the depth [34], and with standard deviation 1 pixel for the image coordinates. The algorithm used for building and merging the maps is very similar to the one in the experiment with real RGB-D data, with the exception that we use the ground-truth data association for the observed features, and the ground truth initial correspondence for the robots. Robots run the method discussed in Sect. 4.3 for 50 steps. During the first 30 steps, they move and build their local maps, and simultaneously, they run the map merging method. During the last 20 steps there are no changes in the local maps, and thus, they run the map merging algorithm to agree on the latest local maps. The robots propagate the changes in their local maps after each 3 steps. Our agents exchange data if they are closer than 3.5 m. Figure 4.1a shows the 3D features position estimated by robot 1 at the last map merging step $k = 50$, for the 100 Monte Carlo simulations. Since the observation noise is small, the points are very accurate and similar to each other. Visually, they are almost indistinguishable.

We have studied the performance of the method using the following metrics [8]: the average root mean square error (RMS); and the average normalized state estimation error squared (NEES). For each robot i, step k, and Monte Carlo simulation l, we let $\tilde{\mathbf{x}}_i^{G,l}(k)$ be the difference between the estimates of the common features' positions in the global map mean $\hat{\mathbf{x}}_i^G(k)$ in Eq. (4.19) and their ground-truth position $\mathbf{x}$. Equivalently, we let $(\hat{I}_i^{A,l}(k+1))^{-1}/d_i(k)$ be the consistent expression of the covariance matrix of the common features, as in Theorem 8, for robot i, step k, and Monte Carlo simulation l. Figures 4.1b, c show the RMS and NEES per step k computed as follows:

$$RMS = \frac{\sqrt{\sum_{l=1}^{100} \sum_{i=1}^{n} \frac{\left(\tilde{\mathbf{x}}_i^{G,l}(k)\right)^T \left(\tilde{\mathbf{x}}_i^{G,l}(k)\right)}{100n}}}{\mathcal{M}}, \qquad (4.40)$$

$$NEES = \sum_{l=1}^{100} \sum_{i=1}^{n} \frac{\left(\tilde{\mathbf{x}}_i^{G,l}(k)\right)^T \left(d_i(k)\hat{I}_i^{A,l}(k+1)\right) \left(\tilde{\mathbf{x}}_i^{G,l}(k)\right)}{100n},$$

where $\mathcal{M}$ is the size of common features.

Figure 4.1b shows the RMS per step (blue solid). Due to the information share, the estimated features positions become more accurate as the iterations go by, reaching estimation errors per coordinate smaller than 1 mm. Figure 4.1c displays the NEES value obtained (red solid), which should follow a χ^2 distribution with $\mathcal{M}$ degrees of freedom. Thus, if the estimated merged maps are consistent, the expected value for the NEES is $\mathcal{M}$ (black solid, dof), and it should not overpass the value $\chi^2_{0.99,dof}$ (black dashed). During all the steps, the estimated features' positions are consistent. This is the expected behavior for systems where robots observe the full 3D position of the features. Recall that we ensure consistency (Theorem 8) as long as the centralized map is consistent, and this depends on the local maps being consistent. Thus, for scenarios where robots only get partial measurements of the features positions, and depending on the particular local mapping method, the local maps may not be consistent. Even in this case, our algorithm will produce estimates more conservative than the centralized map.

4.6 Closure

In this chapter we have presented an algorithm for dynamically merging feature-based maps in a robot network with limited communication. This algorithm allows the robots to have a better map of the environment containing the features observed by any other robot in the team. Thus, it helps the coordination of the team in several multi-robot tasks such as exploration or rescue. The algorithm correctly propagates the new information added by the robots to their local maps. We have shown that, with the studied strategy, the robots correctly track the global map. At the final step, they obtain the last global map, which contains the updated information at all the robots. We have analyzed the performance of the method for robots equipped with RGB-D sensors in a simulated environment. Additional experiments with real data acquired with conventional cameras and with RGB-D sensors, under link failures and switching topologies, are presented in Chap. 5.

References

1. P. Alriksson, A. Rantzer, Distributed Kalman filtering using weighted averaging, in *International Symposium on Mathematical Theory of Networks and Systems*, Kyoto, Japan, July 2006
2. R. Aragues, J. Cortes, C. Sagues, Distributed consensus algorithms for merging feature-based maps with limited communication. Robot. Auton. Syst. **59**(3–4), 163–180 (2011)
3. R. Aragues, J. Cortes, C. Sagues, Distributed consensus on robot networks for dynamically merging feature-based maps. IEEE Trans. Robot. **4**, 850–854 (2012)
4. R. Aragues, J. Cortes, C. Sagues, Distributed map merging with consensus on common information, in *European Control Conference*, Zurich, Switzerland, July 2013, pp. 736–741
5. R. Aragues, C. Sagues, Y. Mezouar, Feature-based map merging with dynamic consensus on information increments, in *IEEE International Conference on Robotics and Automation*, Karlsruhe, Germany, May 2013, pp. 736–741
6. R. Aragues, G. Shi, D.V. Dimarogonas, C. Sagues, K.H. Johansson, Y. Mezouar, Distributed algebraic connectivity estimation for undirected graphs with upper and lower bounds. Automatica **50**, 3253–3259 (2014)
7. R. Aragues, C. Sagues, Y. Mezouar, Feature-based map merging with dynamic consensus on information increments. Auton. Robot. **38**, 243–259 (2015)
8. Y. Bar-Shalom, X. R. Li, T. Kirubarajan, *Estimation with Applications to Tracking and Navigation: Theory Algorithms and Software* (Wiley, New York, 2004)
9. F. Bullo, J. Cortes, S. Martinez, *Distributed Control of Robotic Networks*. Applied Mathematics Series Princeton University Press, Princeton, 2009), http://coordinationbook.info

10. G. Calafiore, Distributed randomized algorithms for probabilistic performance analysis. Syst. Control Lett. **58**(3), 202–212 (2009)
11. G. Calafiore, F. Abrate, Distributed linear estimation over sensor networks. Int. J. Control **82**(5), 868–882 (2009)
12. R. Carli, A. Chiuso, L. Schenato, S. Zampieri, Distributed Kalman filtering based on consensus strategies. IEEE J. Sel. Areas Commun. **26**, 622–633 (2008)
13. D. W. Casbeer, R. Beard, Distributed information filtering using consensus filters, in *American Control Conference*, St. Louis, USA, June 2009, pp. 1882–1887
14. H. Jacky Chang, C.S. George Lee, Y. Charlie Hu, Yung-Hsiang Lu, Multi-robot SLAM with topological/metric maps, in *IEEE/RSJ International Conference on Intelligent Robots and Systems*, San Diego, USA, October 2007, pp. 1467–1472
15. A. Cunningham, V. Indelman, F. Dellaert. DDF–SAM 2.0: consistent distributed smoothing and mapping. In *IEEE International Conference on Robotics and Automation*, Karlsruhe, Germany, May 2013, pp. 5220–5227
16. A. Cunningham, K.M. Wurm, W. Burgard, F. Dellaert, Fully distributed scalable smoothing and mapping with robust multi–robot data association, in *IEEE International Conference on Robotics and Automation*, St. Paul, USA, May 2012, pp. 1093–1100
17. G. Dissanayake, P. Newman, S. Clark, H.F. Durrant-Whyte, M. Csorba, A solution to the simultaneous localization and map building (SLAM) problem. IEEE Trans. Robot. Autom **17**(3), 229–241 (2001)
18. D. Fox, J. Ko, K. Konolige, B. Limketkai, D. Schulz, B. Stewart, Distributed multirobot exploration and mapping. IEEE Proc. **94**(7), 1325–1339 (2006)
19. R.A. Freeman, P. Yang, K.M. Lynch, Stability and convergence properties of dynamic average consensus estimators, in *IEEE Conference on Decision and Control*, San Diego, CA, December 2006, pp. 398–403
20. S. Grime, H.F. Durrant-Whyte, Data fusion in decentralized sensor networks. Control Eng. Pract. **2**(5), 849–863 (1994)
21. R.A. Horn, C.R. Johnson, *Matrix Analysis* (Cambridge University Press, Cambridge, 1985)
22. A. Howard, Multi-robot simultaneous localization and mapping using particle filters. Int. J. Robot. Res. **25**(12), 1243–1256 (2006)
23. G.P. Huang, N. Trawny, A.I. Mourikis, S.I. Roumeliotis, On the consistency of multi-robot cooperative localization, in *Robotics: Science and Systems*, Seattle, WA, USA, June 2009, pp. 65–72
24. G.P. Huang, N. Trawny, A.I. Mourikis, S.I. Roumeliotis, Observability-based consistent EKF estimators for multi-robot cooperative localization. Auton. Robot. **30**(1), 99–122 (2011)
25. S. Huang, Z. Wang, G. Dissanayake, U. Frese, Iterated d-slam map joining: evaluating its performance in terms of consistency, accuracy and efficiency. Auton. Robot. **27**(4), 409–429 (2009)
26. S. Julier, J.K. Uhlmann, General decentralised data fusion with covariance intersection (CI), in *Handbook of Multisensor Data Fusion*, ed. by D.L. Hall, J. Llinas (CRC Press, Boca Raton, 2001)
27. K. Konolige, J. Gutmann, B. Limketkai, Distributed map-making, in *Workshop on Reasoning with Uncertainty in Robotics, International Joint Conference on Artificial Intelligence*, Acapulco, Mexico, 2003
28. A. Leshem, L. Tong, Estimating sensor population via probabilistic sequential polling, IEEE Signal Process. Lett., **12**(5):395–398 (2005)
29. K.Y.K. Leung, T.D. Barfoot, H. Liu, Decentralized localization of sparsely-communicating robot networks: a centralized-equivalent approach. IEEE Trans. Robot. **26**(1), 62–77 (2010)
30. K.Y.K. Leung, T.D. Barfoot, H.H.T. Liu, Decentralized cooperative simultaneous localization and mapping for dynamic and sparse robot networks, in *IEEE/RSJ International Conference on Intelligent Robots and Systems*, Taipei, Taiwan, October 2010, pp. 3554–3561
31. T. Li, J.F. Zhang, Consensus conditions on multi-agent systems with time-varying topologies and stochastic communication noises. IEEE Trans. Autom. Control **55**(9), 2043–2057 (2010)

32. K.M. Lynch, I.B. Schwartz, P. Yang, R.A. Freeman, Decentralized environmental modeling by mobile sensor networks. IEEE Trans. Robot. **24**(3), 710–724 (2008)
33. E.M. Nebot, M. Bozorg, H.F. Durrant-Whyte, Decentralized architecture for asynchronous sensors. Auton. Robot. **6**(2), 147–164 (1999)
34. C.V. Nguyen, S. Izadi, D. Lovell, Modeling kinect sensor noise for improved 3d reconstruction and tracking, in *International Conference on 3D Imaging, Modeling, Processing, Visualization and Transmission*, Zurich, Switzerland, October 2012, pp. 524–530
35. R. Olfati-Saber, Distributed Kalman filter with embedded consensus filters, in *IEEE Conference on Decision and Control* Seville, Spain, 2005, pp. 8179–8184
36. R. Olfati-Saber, Distributed Kalman filtering for sensor networks, in *IEEE Conference on Decision and Control*, New Orleans, LA, December 2007, pp. 5492–5498
37. R. Olfati-Saber, J.S. Shamma, Consensus filters for sensor networks and distributed sensor fusion, *IEEE Conference on Decision and Control* Sevilla, Spain, 2005, pp. 6698–6703
38. L.M. Paz, J.D. Tardos, J. Neira, Divide and conquer: EKF SLAM in $O(n)$. IEEE Trans. Robot. **24**(5), 1107–1120 (2008)
39. M. Pfingsthorn, B. Slamet, A. Visser, A scalable hybrid multi-robot SLAM method for highly detailed maps, in *Lecture Notes in Artificial Intelligence*, ed. by U. Visser, F. Ribeiro, T. Ohashi, F. Dellaert, vol. 5001 (Springer, Berlin, 2008), pp. 457–464
40. W. Ren, R.W. Beard, E.M. Atkins, Information consensus in multivehicle cooperative control. IEEE Control Syst. Mag. **27**(2), 71–82 (2007)
41. D. Rodríguez-Losada, F. Matía, A. Jiménez, Local maps fusion for real time multirobot indoor simultaneous localization and mapping, in *IEEE International Conference on Robotics and Automation*, New Orleans, USA, April 2004, pp. 1308–1313
42. S.I. Roumeliotis, G.A. Bekey, Distributed multirobot localization. IEEE Trans. Robot. Autom. **18**(5), 781–795 (2002)
43. D.P. Spanos, R. Olfati-Saber, R.M. Murray, Distributed sensor fusion using dynamic consensus, in *IFAC World Congress*, Prague, 2005
44. Y.G. Sun, L. Wang, G. Xie, Average consensus in networks of dynamic agents with switching topologies and multiple time-varying delays. Syst. Control Lett. **57**(2), 175–183 (2008)
45. S. Thrun, Y. Liu, D. Koller, A. Ng, H. Durrant-Whyte, Simultaneous localisation and mapping with sparse extended information filters. Int. J. Robot. Res. **23**(7–8), 693–716 (2004)
46. S. Utete, H.F. Durrant-Whyte, Routing for reliability in decentralised sensing networks. Am. Control Conf. **2**, 2268–2272 (1994)
47. D. Varagnolo, G. Pillonetto, L. Schenato, Distributed statistical estimation of the number of nodes in sensor networks, in *IEEE Conference on Decision and Control* Atlanta, USA, 2010, pp. 1498–1503
48. R. Vincent, D. Fox, J. Ko, K. Konolige, B. Limketkai, B. Morisset, C. Ortiz, D. Schulz, B. Stewart, Distributed multirobot exploration, mapping, and task allocation. Ann. Math. Artif. Intell. **52**(1), 229–255 (2008)
49. S.B. Williams, H.Durrant-Whyte, Towards multi-vehicle simultaneous localisation and mapping, in *IEEE International Conference on Robotics and Automation*, Washington, DC, USA, May 2002, pp. 2743–2748
50. L. Xiao, S. Boyd, Fast linear iterations for distributed averaging. Syst. Control Lett. **53**, 65–78 (2004)
51. L. Xiao, S. Boyd, S. Lall, A space-time diffusion scheme for peer-to-peer least-square estimation, in *Symposium on Information Processing of Sensor Networks (IPSN)*, Nashville, TN, April 2006, pp. 168–176
52. X.S. Zhou, S.I. Roumeliotis, Multi-robot SLAM with unknown initial correspondence: the robot rendezvous case, in *IEEE/RSJ International Conference on Intelligent Robots and Systems*, Beijing, China, October 2006, pp. 1785–1792
53. M. Zhu, S. Martínez, Discrete-time dynamic average consensus. Automatica **46**(2), 322–329 (2010)

Chapter 5
Real Experiments

Abstract We show some experiments of the methods studied in this book with real data under different communication schemes. We have carried out experiments using a data set from Frese and Kurlbaum, a data set for data association, 2008, [1] with bearing-only information extracted from conventional images. Additionally, we have analyzed the performance of the data association method and the localization and map merging algorithms under real data acquired with an RGB-D sensor, which provides both visual and depth information.

Keywords RGB-D data · Visual data · Distributed and parallel algorithms · Localization · Data association · Map merging

5.1 Data Association with Visual Data

The behavior of the data association algorithm is analyzed with the data set [1] with bearing information obtained with vision (Sony EVI-371DG), in an environment of 60 m × 45 m performing 3297 steps. It is an indoor scenario where the robot moves along corridors and rooms. The data set contains real odometry data and images captured at every step (Fig. 5.1). The images are processed and measurements to natural landmarks are provided. The natural landmarks are vertical lines extracted from the images and processed in the form of bearing-only data. The observations in the dataset are labeled so that we have the ground truth data association. This dataset is very challenging for a conventional visual map building algorithm due to the limited field of view of the camera (Sony EVI-371DG). Furthermore, the camera is pointing forward in the same direction of robot motion and the robot traverses rooms and corridors with few features in common. Note that this situation is much more complex than situations where the camera can achieve big parallax, or systems with ominidirectional cameras, where features within 360° around the robot are observed. We analyze the performance of the algorithm under 3 communication graphs (Fig. 5.2a–c). We select 9 subsections of the whole path for the operation of 9 different robots (Fig. 5.2d). A separate SLAM is executed on each subsection, producing the 9 local maps (Fig. 5.2e). The local data associations are computed

© The Author(s) 2015
R. Aragues et al., *Parallel and Distributed Map Merging and Localization*,
SpringerBriefs in Computer Science, DOI 10.1007/978-3-319-25886-7_5

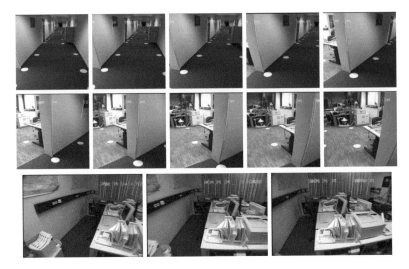

Fig. 5.1 An example of the images used by the 9 robots to test the proposed method [1]. Although the data set also provides artificial landmarks (*white circles* on the floor), we do not use them, and instead we test the algorithm using the lines extracted from natural landmarks (in *yellow*)

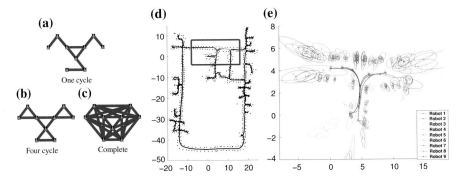

Fig. 5.2 **a–c** Communication graphs. **d** Section of the dataset used in the experiments and trajectories followed by the 9 robots. **e** Local maps acquired by the 9 robots

using the JCBB [5] since it is very convenient for clutter situations like the considered scenario (Fig. 5.2e).

Table 5.1 gives statistics about the number of inconsistencies found considering the different network topologies in Fig. 5.2a–c. We show the obtained associations compared to the ground truth results. The number of association sets is the number of connected components of $\mathbf{A}^t$. The number of good links (true positives) is obtained associations between 2 features which are true (ground truth). The missing links (false negatives) are associations that are in the ground truth information, but have been not detected. And spurious links (false positives) are associations found between features that are different according to the ground truth. The sixth row, C, is the number of

Table 5.1 Initial associations between the 9 local maps

Comm. graph	(a)	(b)	(c)
Association sets (ground truth)	242	284	400
Association sets	182	218	290
Good links (true positives)	160	190	228
Missing links (false negatives)	82	94	172
Spurious links (false positives)	22	28	62
Conflictive sets (C)	3	5	8
Number of features m_{sum}	138	144	154
Conflictive features	16	24	51

Table 5.2 Management of the inconsistencies

Comm. graph	(a)	(b)	(c)
Iterations	1	1	1
Initial conflictive sets	3	5	8
Resulting conflictive sets	0	0	0
Resulting incons. feats.	0	0	0
Deleted links	6	10	34
Good deleted links (true positives)	2	2	12
Spurious deleted links (false positives)	4	8	22

conflictive sets. The next row in the table shows the total number of features which has been associated to any other feature from other local map. The last row gives information about how many of those features are conflictive. The amount of missing and spurious associations obtained is very high for the three network topologies. This is the expected result for many real scenarios, where the landmarks are close to each other, and where the only available information is their cartesian coordinates. As a result, the conflictive features are more than a 10 % of the total. In communication graphs with more cycles (Fig. 5.2b, c), there are more conflictive features. In the three cases, after a single execution of the detection and the resolution algorithms, all the inconsistencies are solved (Table 5.2, 1st row). An interesting result is that, although the algorithm presented in Chap. 2 cannot distinguish between good and spurious edges, in practice a high number of the deleted edges (last row) are spurious.

5.2 Map Merging with Visual Data

We also show results of some experiments of the map merging method using real data from the previously described data set [1] with bearing information obtained with vision (Sony EVI-371DG). The landmarks are vertical lines extracted from the images (Fig. 5.1). We have carried out these experiments with 9 robots. The total area covered by the robots is a square of $30\,\mathrm{m} \times 30\,\mathrm{m}$ (Fig. 5.3a). We run a separate SLAM in each robot and obtain 9 maps. We use a bearing-only SLAM algorithm with features parameterized in inverse depth [3] followed by a transform to Cartesian coordinates before the merging process. We express the local maps in global coordinates according to the relative robot poses seen in Fig. 5.3a, obtaining the results shown in Fig. 5.3b. Note that this is the result of putting the maps together, without applying a merging method. The team of robots build local maps of the environment, and then they execute the fusion algorithm presented in Chap. 4 to merge these maps.

   We study the behavior of the map merging method under three different scenarios: a fixed communication graph, a graph with switching topology, and a graph with link failures (Fig. 5.4).

   We illustrate the performance of our algorithm by comparing the global map estimated by the robots along the iterations with the actual global map. We consider the x-coordinate of feature F23 (within the black box in Fig. 5.3). In Fig. 5.5 we show the estimated information matrices $\hat{I}_i^G(k)$ and vectors $\mathbf{i}_i^G(k)$ (solid lines) during 40 iterations, compared to the global map $I_G^k, \mathbf{i}_G^k$, Eq. (4.8) (thick solid line). In all cases the estimates converge to the average value very fast. Note that at iteration 0, robots 1, 7, 8 and 9 have estimated different initial values for F23. Then, they execute the map merging method proposed and they reach an agreement.

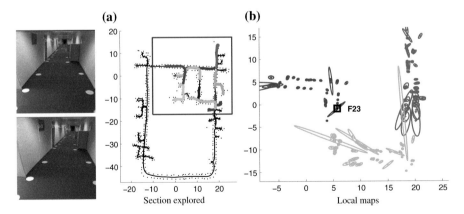

(a)   **(b)**

Section explored   Local maps

Fig. 5.3 **a** Trajectories followed by the 9 robots. They cover a region of $30\,\mathrm{m} \times 30\,\mathrm{m}$ of the data set. **b** Local maps obtained by robots 2 (*green*), 6 (*yellow*), and 9 (*pink*). The feature F23 within the *black box* will be used for testing purposes within this section

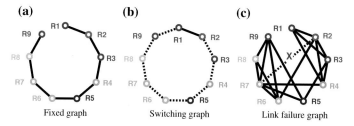

Fig. 5.4 Communication graphs. **a** Fixed string graph, with robots 1 and 9 in the extremes. **b** For each iteration k, there exists a single edge linking robots $((k-1) \mod 9) + 1$ and $(k \mod 9) + 1$. **c** A connected graph where at each iteration one of its link fails

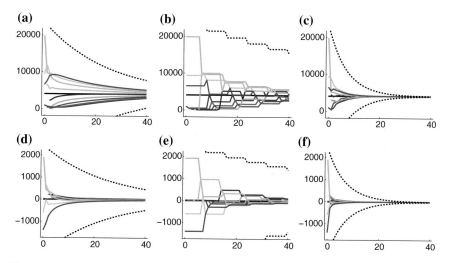

Fig. 5.5 Estimated position (x-coordinate) of F23 at each robot along 40 iterations. We display its associated components within the information matrices $\hat{I}_i^G(k)$ (*first row*) and vectors $\hat{i}_i^G(k)$ (*second row*). We analyze the results for the graphs in Fig. 5.4. **a** $\hat{I}_i^G(k)$ Fixed graph. **b** $\hat{I}_i^G(k)$ Switching graph. **c** $\hat{I}_i^G(k)$ Link failure graph. **d** $\hat{i}_i^G(k)$ Fixed graph. **e** $\hat{i}_i^G(k)$ Switching graph. **f** $\hat{i}_i^G(k)$ Link failure graph

Figure 5.6 analyzes the evolution of the mean $\hat{\mathbf{x}}_i^G(k)$ and covariance $\hat{\Sigma}_i^G(k)$ (Chap. 4) estimated by each robot i (solid lines) for feature F23, x-coordinate. It can be seen that $\hat{\mathbf{x}}_i^G(k)$ converges to $\hat{\mathbf{x}}_G^k$, Eq. (4.5), (thick solid), and that the consistent expression of the covariance matrix $(\hat{I}_i^A(k+1))^{-1}/d_i(k)$ in Theorem 8 remains larger than the centralized covariance (thick solid) for all robots and all steps. We also display the numerical covariance, which cannot be locally computed by the robots, $E[(\hat{\mathbf{x}}_i^G(k) - \mathbf{x})(\hat{\mathbf{x}}_i^G(k) - \mathbf{x})^T]$ (dashed lines), which converges to the centralized covariance.

We analyze the effects of the communication topology on the performance of the algorithm. In the fixed and the switching communication graphs (Fig. 5.5, first and second column), the convergence is slower than for the link failure graph

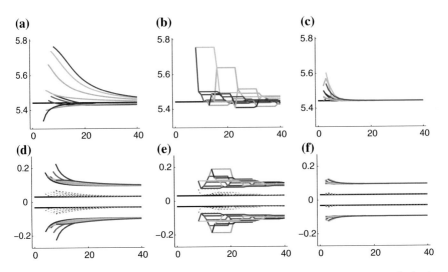

Fig. 5.6 Estimated position (x-coordinate) of F23 at each robot along 40 iterations. We display its associated components within the mean $\hat{\mathbf{x}}_i^G(k)$ (*first row*) and covariance $\hat{\Sigma}_i^G(k)$ (*second row*) using *color solid lines*. We analyze the results for the graphs in Fig. 5.4. **a** $\hat{\mathbf{x}}_i^G(k)$ Fixed graph. **b** $\hat{\mathbf{x}}_i^G(k)$ Switching graph. **c** $\hat{\mathbf{x}}_i^G(k)$ Link failure graph. **d** $\hat{\Sigma}_i^G(k)$ Fixed graph. **e** $\hat{\Sigma}_i^G(k)$ Switching graph. **f** $\hat{\Sigma}_i^G(k)$ Link failure graph

(Fig. 5.5, third column). In this fixed graph (Fig. 5.4a) the topology is a string, with robots 1 and 9 in the extremes. This is a specially bad configuration since the time needed to propagate information from the extreme robots to the whole network is maximal. The per step convergence factor $\gamma = |\lambda_2(W)|$ (4.36) depends on the Metropolis weights matrix, which is

$$W = \frac{1}{3} \begin{bmatrix} 2 & 1 & 0 & \dots & 0 \\ 1 & 1 & 1 & 0 & 0 \\ 0 & \ddots & \ddots & \ddots & 0 \\ \vdots & \mathbf{0} & 1 & 1 & 1 \\ 0 & 0 & 0 & 1 & 2 \end{bmatrix}.$$

We obtain a value for $\gamma = 0.96$ close to 1. This produces a slow convergence. The convergence bounds are displayed in black dashed lines (Fig. 5.5, first column). In the switching graph case (Fig. 5.4b), at every time instant, only one communication link exists in the graph and this sequence takes place in a circular fashion. This is a very extreme communication scheme where, although the conditions for convergence are satisfied, the converge speed is expected to be slow. We can see that (Fig. 5.5, second column) for each robot, estimates remain unchanged during long periods of time, then they experiment two consecutive changes, and then they remain unchanged again. Each robot remains isolated during 7 iterations, maintaining its estimates unchanged.

Then, it exchanges information with its previous neighbor and, in the next iteration, with its next neighbor. In our case, at each iteration k, there exists a single edge linking robots $((k - 1) \mod 9) + 1$ and $(k \mod 9) + 1$. The index of joint connectivity is $\tau = 8$ since every 8 iterations the joint graph is connected. There are only 9 different Metropolis weight matrices $W(k)$, depending on the linked robots at time k, that are repeated successively. We obtained a value for $\delta = 0.89$ using (4.37). We draw the bounds using black dashed lines (Fig. 5.5, second column).

In the link failure graph (Fig. 5.4c), at each iteration one of the links in the graph fails although the graph remains connected. Thus, we obtain an index of joint connectivity of $\tau = 1$. Evaluating all the possible Metropolis weight matrices in this graph, we obtain $\delta = 0.80$. We show the convergence speed bounds (Fig. 5.5, third column) using black dashed lines. This communication scheme exhibits the fastest convergence speed, since $0.80^k \leq 0.96^k \leq 0.89^{\lfloor k/8 \rfloor}$ for all $k = 0, 1, \ldots$ This faster convergence can also be observed in the estimated mean and covariance (Fig. 5.6), where the estimates approach the global map faster for the link failure graph (third column). It is noted that regardless of the presence of link failures or changes in the communication topology, the numerical covariance remains bounded by the locally computed covariance matrix.

In addition, we display (Fig. 5.7) the global map estimated by robot 1 after 5, and 20 iterations (colored lines) of the merging algorithm, and under the fixed communication graph (Fig. 5.4a). The maps estimated by the 9 robots are similar. We compare the estimates at robot 1 to the global map in (4.5) (black lines). Due to the network configuration, after 5 iterations robot 1 has received information from the initial local maps of robots 1–6. However, it still knows nothing of the local maps of robots 7–9 (Fig. 5.7a). As previously stated, this fixed communication graph has a slow convergence speed. However, after 20 iterations the map estimated by robot 1 is very close to the global map (Fig. 5.7b). In addition, it is observed that the information fusion leads to a great improvement in the map quality, where not only the uncertainty is greatly decreased, but also the local maps are corrected.

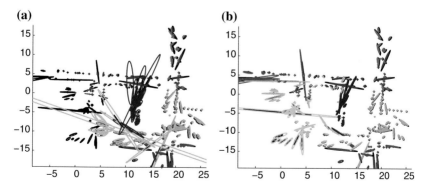

Fig. 5.7 Global map estimated by robot 1 after 5 (**a**) and 20 (**b**) iterations of the merging algorithm, and under the fixed graph (Fig. 5.4a). Different *colors* identify the source local map. Although the global map contains a single estimate per feature, the features observed by more than one robot are displayed by multiple *colored ellipses*. The global map $\hat{\mathbf{x}}_G^k$, Σ_G^k is displayed in *black*

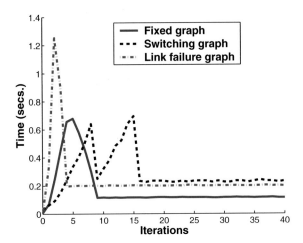

Fig. 5.8 Execution times (per iteration and robot) exhibited by the merging algorithm

Finally, we show the performance of the map merging algorithm in terms of execution times (Fig. 5.8). During the first iterations, the peaks on the execution times are due to the expansion and arrangement of the information matrices and vectors $\hat{I}_i^G(k)$ and $\hat{i}_i^G(k)$ that are performed by the robots whenever they discover new features in its neighbors' information. These memory allocation operations, which are computationally expensive, give rise to this behavior. In the fixed graph case (in blue solid) this situation continues until iteration 5, when robot 5, the robot in the central position within the string graph, has received information from all the robots. Its information matrix $\hat{I}_i^G(5)$ reaches its maximum size and, from here to the end of the experiment, its global map estimate changes but its size remains unchanged. The execution times reach a peak at iteration 5 and from here on, it decreases. From iterations 5 to 9 other robots achieve the maximal size of their matrices $\hat{I}_i^G(k)$, and finally, from iteration 9 to the end of the experiment, the global map size remains unchanged for all the robots. For this reason, we can see that the execution times are drastically reduced from iteration 9 to the end of the experiment. For the switching graph (in black dashed), the first robots that receive information from all the others are 8 and 9, at iteration 8, and from iterations 9 to 15 robots 1–7 successively expand their maps to the maximum size. Finally, from iteration 16 to the end, all the robot's global map estimates present the maximal size and only its contents change. For this reason, the execution times decrease. Finally, for the link failure graph (in dotted red), due to its higher connectivity, all the expansion operations are carried out duringthe first 4 iterations, giving rise to the larger peak

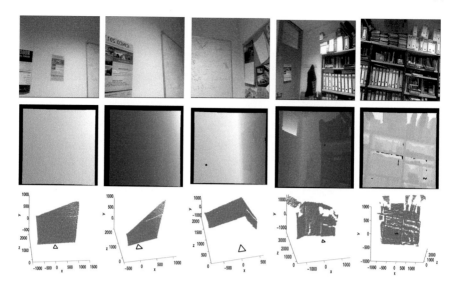

Fig. 5.9 An example of the images obtained by the 9 robots with the RGB-D sensor

observed in the plot. After these expansion operations, the execution times are similar for the three communication graphs.

5.3 Data Association, Localization, and Map Merging with RGB-D

We show experiments using RGB-D cameras, which provide both regular RGB (Fig. 5.9, first row) and depth image information (Fig. 5.9, second row). Thus, it is possible to compute the cloud of points in 3D from a single image (Fig. 5.9, third row). We consider a scenario with 9 robots. Initially, robots are placed at unknown poses in the environment. From their initial pose, robots take an image of the scene (Fig. 5.10). They extract SIFT/SURF features [2] from their RGB images and they use the depth information and the camera parameters to compute the 3D position of these features, as in Fig. 5.9, third row.

Given the sets of features of two robots, it is possible to establish matches based on the SIFT/SURF descriptor of the features. Then, robots compute their relative pose (rotation and translation) using the matches and the 3D position of the features in a robust way (RANSAC) and discard matches that disagree with the most supported relative pose candidate (Fig. 5.11). Typically, the number of matches between overlapping images is high, and the noise in the image points is small. Therefore, the previous method provides highly accurate results. Robots use their initial images to compute the relative poses of their nearby robots (Fig. 5.10) and based on this

Fig. 5.10 Images taken from the first robot poses. We are depicting the matches (*red lines*) between some of the images

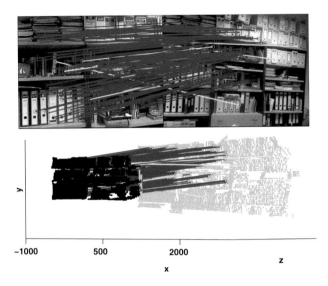

Fig. 5.11 SURF/SIFT matches (*red*) that satisfy the relative rotation and translation restriction. The ones that are rejected are depicted in *yellow*

information, they compute their pose in the common reference frame as explained in Chap. 3. This method to obtain the robot poses in the common frame is not only restricted to RGB-D images taken from the first robot poses. It can be equivalently applied to images acquired during the exploration, or to the local maps of the robots.

Each robot explores a region and builds a map of the environment. We let each robot execute a SLAM algorithm with SIFT/SURF features parameterized in 3D cartesian coordinates. Robots are represented by their 3D position and orientation, and robot motions are predicted by computing the relative rotation and translation between successive images. Figure 5.12 shows the resulting map (red points and ellipses) obtained by robot 3 (dark gray triangle) along its trajectory (dashed line). We also show the 3D RGB-D points observed from some of the steps of the robot trajectory (light gray points) to give an idea of the scene. After exploring, nearby robots compute the local data association between their maps based on the SIFT/SURF de-

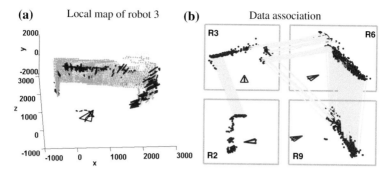

Fig. 5.12 **a** Local map (*red points* and *ellipses*) of robot 3 (*triangle*). The RBG-D points observed from some steps of the robot trajectory are displayed (*gray points*) to give an idea of the scene. **b** Robots (*triangles*) compute the data association (*lines*) between their local maps (*dark dots*). We display the associations between robots R2, R3, R6, and R9

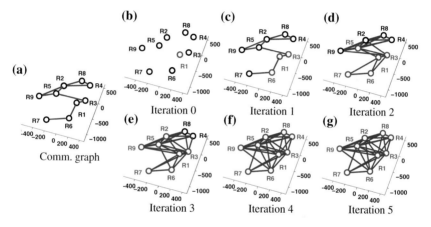

Fig. 5.13 **a** Robots (*circles*) exchange data with their neighbors in the communication graph (linked trough *lines*). *Blue lines* in figures **b–g** indicate that the two robots have received information from each other during the previous iterations. *Red circles* are robots that have received data from R1; equivalently, R1 has received data of these robots as well

scriptors and the position of their features. Then, they propagate the local associations and find and solve inconsistencies as explained in Chap. 2 (Fig. 5.12).

Finally, they merge their local maps and build a global map of the environment using the communication graph in Fig. 5.13a. In Fig. 5.13b–g we explain how this communication graph affects the information exchange. We are displaying a link (blue line) between pairs of robots that have received information from each other during the previous iterations. Initially (Fig. 5.13b), each robot only has its local information and thus there are no lines in the graph. During the first iteration (Fig. 5.13c), robots exchange information with their one-hop neighbors (blue lines in Fig. 5.13a). At iteration 2 robots exchange data with their neighbors again according to the graph

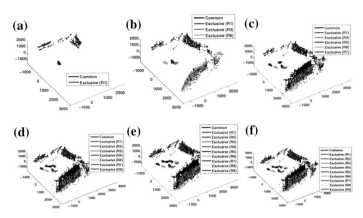

Fig. 5.14 Global map estimated by robot $R1$ at iterations 0–5. Common features observed by several robots and exclusive areas observed by a single robot are depicted in different *colors*. **a** Iteration 0, **b** Iteration 1, **c** Iteration 2, **d** Iteration 3, **e** Iteration 4, **f** Iteration 5

in Fig. 5.13a, and thus they have access to two-hop neighbors data (lines in Fig. 5.13d). The process is repeated during the next iteration, having access to three-hop neighbors data (Fig. 5.13e), and so on. After iteration 4 (Fig. 5.13f, g), each robot has received information from all the other robots. Additional iterations allow the robots to obtain a more accurate estimate of the global map.

Robots fuse their maps as explained in Chap. 4. Figure 5.14 shows the global map estimated by robot $R1$ along iterations 0–4, with contains features observed exclusively by a single robot (exclusive), as well as features observed by several robots (common). After 4 iterations, robot $R1$ has received information from all the other robots and thus its map already contains estimates for all the features observed by the team. Successive iterations of the map merging algorithm produce more accurate estimates of the features (Fig. 5.15).

In the second set of experiments, we study a scenario that is more challenging from the point of view of the map merging method (Chap. 4). We consider again a robot team composed by 9 robots that acquire information with RGB-D sensors, and that extract SIFT features [2] from the images. The robots take 473 images in total, and from each image around 1333 SIFT points are extracted. This time, robots do not merge their maps after exploring the region, but we consider different situations. Four of the robots ($R3$, $R5$, $R7$, $R9$) have already finished their exploration when the merging process begins; they provide their local maps at the step $k = 0$ and remain static during the execution of the algorithm. Robots $R2$, $R6$, $R8$ on the other hand, keep on moving and updating their local maps simultaneously to the merging process. Finally, robots $R1$ and $R4$ explore and update their maps as well, but they form a different exploration cluster and remain disconnected from the team for several steps. A summary of the time steps when robots propagated their local maps in our experiment can be seen in Table 5.3. The local maps of the robots contain around 962 features per map at the last step; the smallest and largest local maps belong to robots $R2$ and $R9$ and have respectively 163 and 2858 features. We solve both initialization

Table 5.3 Steps $\mathscr{T}_i$ at which robot i propagates its local map

Fixed agents	Exploring agents	Other cluster
$\mathscr{T}_3 = \{0\}$		
$\mathscr{T}_5 = \{0\}$	$\mathscr{T}_2 = \{0, 4, 8\}$	
$\mathscr{T}_7 = \{0\}$	$\mathscr{T}_6 = \{0, 5, 10, 20\}$	$\mathscr{T}_1 = \{5, 15, 25\}$
$\mathscr{T}_9 = \{0\}$	$\mathscr{T}_8 = \{0, 5, 10, 20\}$	$\mathscr{T}_4 = \{5, 15, 25\}$

and data association in a centralized fashion, using the same method explained for the first experiment, and focus on the behavior of the map merging method.

As robots move, the communication graph $\mathscr{G}_k$ changes and new links appear and disappear (Fig. 5.16); for instance, $R2$ gets isolated for some steps ($k = 6$); $R1$ and $R4$ remain isolated from the others ($k = 0$, $k = 6$) until step $k = 33$; and the neighbors of all the robots change several times ($k = 0$ to $k = 40$). Note that in none step it is a complete (all-to-all) graph.

We show the evolution of the covariances and mean vectors, and information matrices and vectors of the global map estimated by the robots (Fig. 5.17). We illustrate it using the x-coordinate of a feature $F_{2,31}$ which was observed by robots in the cluster ($R1$, $R4$), and in the remaining team ($R2$, $R3$, $R5$, $R6$, $R7$, $R8$, $R9$). At each step, we display (blue solid) the estimate that would be obtained by a centralized system (Eq. (4.5)) considering all the robot local maps. Note that the centralized estimates change whenever a robot propagates changes of its local map (Table 5.3). The mean $\hat{\mathbf{x}}_i^G(k)$, covariance $\hat{\Sigma}_i^G(k)$, and information matrix $\hat{I}_i^G(k)$ and vector $\mathbf{i}_i^G(k)$ estimated by all the robots (different colors, dashed) correctly converge to the centralized value (blue solid). Note that the covariance estimates (Fig. 5.17b, different colors, dashed) can become smaller than the global one (blue solid) for some robots and iterations, whereas the consistent expression of the covariance matrix $(\hat{I}_i^A(k+1))^{-1}/d_i(k)$ in Theorem 8 (Fig. 5.17c, different colors, dashed) remains larger than the central-

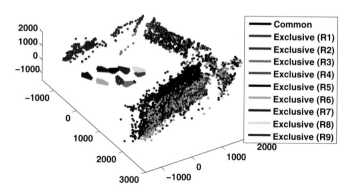

Fig. 5.15 Localization of the robots and global map estimated by robot $R1$ after 20 iterations of the map merging algorithm

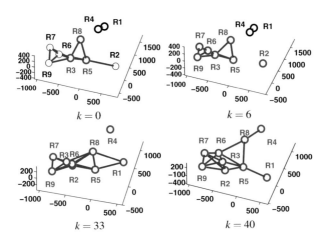

Fig. 5.16 Communication graphs $\mathcal{G}_k$ at different steps k. Robot $R8$ has received information of the local maps of the robots displayed in *red*. x-, y- and z- axes in are in millimeters

ized covariance (blue solid) for all robots and all steps. Since up to step $k = 33$ robots remain divided into two different clusters, we show as well (green solid) the estimate that would be obtained by a centralized system (Eq. (4.5)), but considering *only* the robot local maps in each cluster. During the time both clusters are separated, the estimates of different robots (different colors, dashed), correctly track this cluster centralized value (green solid) that contains all the information that could be available in the best case to the robots. The robot estimates react to changes in the local maps in an appropriate way. In particular, up to iteration $k = 33$, since the cluster composed by $R1$, $R4$ has a complete (all-to-all) topology, their estimates are exactly equal to the cluster centralized ones (green solid).

We make an analysis of the communication and memory costs of our algorithm (Fig. 5.18, left column a, c, e). These cost include both the consensus on the common features, as well as the propagation of the mean and the elements in the main diagonal of the covariance matrix for the exclusive features and robot poses (Sect. 4.4). These exclusive features and robot poses are re-estimated at each step based on the most recent estimates of the common features. We consider numbers encoded with single precision (4 bytes). A benefit of using a consensus-based algorithm is its low memory cost (Fig. 5.18e) of around 45 MBytes per robot, which does not depend on the number of robots but only on the scene size. In addition, the communication cost per iteration (Fig. 5.18a, c) is almost the same for all the robots; observe that there are almost no differences between the average (gray solid) and maximum costs (black dashed). We have compared our performance against a method based on propagation (Fig. 5.18, right column b, d, f). The memory usage (Fig. 5.18f) of the propagation method is much higher than for our method (Fig. 5.18e). If we sum up the average communication costs per robot (Fig. 5.18 a–d, gray solid line) for the 45 iterations (sum of the along the x-axis), we obtain a total of 234 MBytes

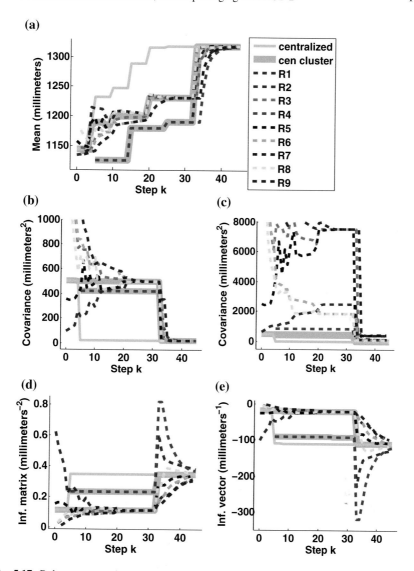

Fig. 5.17 Robots execute the algorithm for fusing their maps for 45 iterations k (x-axis). We show the evolution of the estimates at each robot (different *colors*, *dashed*) of: **a** the mean vector $\hat{\mathbf{x}}_i^G(k)$; **b** the covariance matrix $\hat{\Sigma}_i^G(k)$; **c** the consistent expression for the covariance matrix, $(\hat{I}_i^A(k+1))^{-1}/d_i(k)$ (Theorem 8); **d** the information matrix $\hat{I}_i^G(k)$; and **e** the information vector $\hat{\mathbf{i}}_i^G(k)$. We focus on the evolution of the entry associated to the x-coordinate of feature $F_{2,31}$. We display in *blue solid* the value of this feature coordinate in the global map (Eq. (4.5)). Until step $k = 33$, robots remain in two separated clusters, one of them composed by $R1$, $R4$, and the other by the remaining robots. We display as well (*green solid*) the centralized map that would be obtained by considering all the available local maps within each cluster. After step $k = 33$, both cluster global maps (*green solid*) become the equal to the global map (Eq. (4.5)) that considers the local maps of all the robots (*blue solid*)

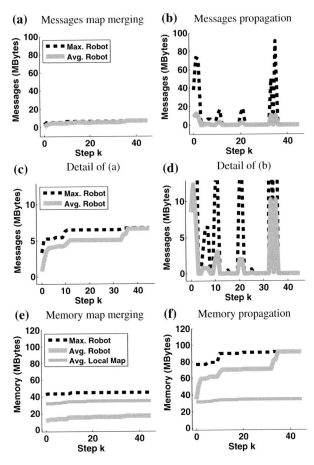

Fig. 5.18 Messages exchanged (**a**)–(**d**) and memory usage (**e, f**) per robot along 45 iterations of our algorithm (*left* column) against a simple propagation method (*right* column). Numbers are encoded with single precision (4 bytes). We show the average amount of information (*gray solid*) as well the largest amount of information per robot (*black dashed*). Figures (**c, d**) show a detail of (**a, b**). In figures exhibiting memory costs (**e, f**), we show the average memory used by the local maps (*green solid*)

versus the 61 MBytes used by the propagation method. This means that, due to the iterative nature of our algorithm, we obtain a total communication cost larger than for the propagation method. However, paying attention to the communication costs per robot (black dashed), in our method all the robots exchange similar amounts of data (Fig. 5.18a, c), whereas the propagation solution exhibits large communication cost peaks (Fig. 5.18b, d). If robots propagated their observations, i.e., the 3D SIFT point clouds extracted from their images, then the costs per robot up to step $k = 33$ would be 322 MBytes storage and 286 MBytes communication. Obviously, propagating the raw RGB + depth images instead is an even worse option; the costs per robot up to step $k = 33$ would be 693 MBytes storage and 616 MBytes communication. After propagating the observations, one of the robots would compute and propagate the global map, with an associated extra cost. Thus, as it can be seen, propagating measurements is not efficient, and propagating local maps is memory demanding and it is prone to large peaks in the communication costs.

Note that the communication costs in Fig. 5.18 do not include the data association. This cost is highly dependent on the method used to match the features. A deep discussion of the performance of different matching strategies can be found in [4].

5.4 Closure

In this chapter we have presented several experiments with real data that confirm the performance of the Map Merging and Localization methods presented in the book. It has been shown that their flexibility allows the robots to easily cope with situations of real-world scenarios, where communication is limited and robots get disconnected from the team for long periods of time. The theoretical results are shown to be correct, and the robots appropriately track the merged map. As discussed, distributed methods have low memory complexity. Due to their iterative nature, if we sum up the communication complexity along all the iterations, they have higher communication consumption than alternative methods based on propagation. However, the communication complexity remains almost constant at each iteration, and they do not generate large communication peaks in some areas of the network. An interesting future area of research directed toward reducing the communication cost would be to let the robots decide when they can stop performing additional merging iterations.

References

1. U. Frese, J. Kurlbaum, A data set for data association. Electronically, (2008) http://www.sfbtr8. spatial-cognition.de/insidedataassociation/
2. D.G. Lowe, Object recognition from local scale-invariant features. IEEE Int. Conf. Comput. Vis. **709**, 1150–1157 (1999)
3. J.M.M. Montiel, J. Civera, J. Davison, Unified inverse depth parametrization for monocular SLAM, in *Robotics: Science and Systems*, Philadelphia, USA, August 2006
4. E. Montijano, R. Aragues, C. Sagues, Distributed data association in robotic networks with cameras and limited communications. IEEE Trans. Robot. **29**(6), 1408–1423 (2013)
5. J. Neira, J.D. Tardós, Data association in stochastic mapping using the joint compatibility test. IEEE Trans. Robot. Autom. **17**(6), 890–897 (2001)

Chapter 6
Conclusions

Keywords Networked robots · Distributed systems · Parallel computation · Limited communication · Multi-robot perception · Localization · Data association · Map merging

Along this book, we have presented distributed methods for localization and merging stochastic feature-based maps acquired by a team of robots for scenarios with limited communication.

In the first place, we have studied a very important problem which appears in most localization and map merging scenarios, and which is particularly hard in multi-robot systems: The association of the features observed by the different robots. We have presented a distributed technique to match several sets of features observed by a team of robots in a consistent way under limited communications. Local associations are found only within robots that are neighbors in the communication graph. After that, a fully decentralized method to compute all the paths between local associations is carried out, allowing the robots to obtain the relationship between their own features and the ones observed by the other team members. In addition, each robot detects all the inconsistencies related with their observations. For every conflictive set detected, in the second step the method is able to delete local associations to break the conflict using only local communications. The whole method is proved to finish in finite time finding and solving all the inconsistent associations.

The problem of localization travels together with the map merging problem. In multi-robot systems, the establishment of a common reference frame is, in addition, very important. Usually, robots start at unknown poses and do not share any reference frame. The localization problem consists of establishing this common frame and computing the robots' poses. Each robot is capable of measuring the relative pose of its neighboring robots. However, it does not know the poses of far robots, and it can only exchange data using the range-limited communication network. The network localization problem has been studied for different scenarios: the planar pose network localization from noisy relative measurements relative to an anchor node; and the position network computation for higher dimensional scenarios, from noisy measurements with simultaneous estimation of the centroid. We have analyzed distributed strategies that allow the robots to agree on a common global frame,

© The Author(s) 2015
R. Aragues et al., *Parallel and Distributed Map Merging and Localization*,
SpringerBriefs in Computer Science, DOI 10.1007/978-3-319-25886-7_6

and to compute their poses or positions relative to the global frame. The presented algorithms exclusively rely on local computations and data exchange with direct neighbors and have been proved to converge under mild conditions on the communication graph. Besides, they only require each robot to maintain an estimate of its own pose. Thus, the memory load of the algorithm is low compared to methods where each robot must also estimate the positions or poses of any other robot.

The robots explore an environment and build their local maps. Simultaneously, the robots communicate and build a global map of the environment. We have deeply discussed this map merging problem and have proposed distributed solutions. The methods studied are fully distributed, relying exclusively on local interactions between neighboring robots. Under fixed connected communication graphs, or time-varying jointly connected topologies, the estimates at each robot asymptotically converge to the global map. Moreover, the intermediate estimates at each robot present interesting properties that allow their use at any time: the mean of the global map estimated by each robot is unbiased at each iteration, and the covariance of the global map estimated by each robot is bounded by the locally computed covariance. The robustness of the map fusion algorithm under link failures and changes in the communication topology has been analyzed theoretically and tested experimentally. The algorithm allows the robots to have a better map of the environment containing the features observed by any other robot in the team. Thus, it helps the coordination of the team in several multi-robot tasks such as exploration or rescue.

Experimental results show the performance of the presented algorithms. We have included simulations at the end of each chapter to show the performance of the studied methods against the known ground truth and we have also included a chapter with real experiments using vision information of conventional cameras and of RGB-D cameras.

Appendix A
Averaging Algorithms and Metropolis Weights

Throughout this document, we frequently refer to averaging algorithms. They have become very popular in sensor networks due to their capability to reach agreement in a distributed way. Let us assume that each robot $i \in \mathcal{V}$ has initially a scalar value $z_i(0) \in \mathbb{R}$. Let $W \in \mathbb{R}_{\geq 0}^{n \times n}$ be a doubly stochastic matrix such that $W_{i,j} > 0$ if $(i,j) \in \mathcal{E}$ and $W_{i,j} = 0$ when $j \notin \mathcal{N}_i$. This matrix is such that $W_{i,i} \in [\alpha, 1]$, $W_{i,j} \in \{0\} \cup [\alpha, 1]$ for all $i,j \in \mathcal{V}$, for some $\alpha \in (0, 1]$. Assume the communication graph $\mathcal{G}$ is connected. If each robot $i \in \mathcal{V}$ updates $z_i(t)$ at each time step $t \geq 0$ with the following averaging algorithm,

$$z_i(t+1) = \sum_{j=1}^{n} W_{i,j}\, z_j(t), \tag{A.1}$$

then, as $t \rightarrow \infty$, the variables $z_i(t)$ reach the same value for all $i \in \mathcal{V}$, i.e., they reach a consensus. Moreover, the consensus value is the average of the initial values,

$$\lim_{t \rightarrow \infty} z_i(t) = z_\star = \frac{1}{n} \sum_{j=1}^{n} z_j(0), \tag{A.2}$$

for all $i \in \mathcal{V}$ [1, 2]. Observe that each robot i updates its variables $z_i(t)$ using local information since the weight matrix has zero entries for nonneighboring robots, $W_{i,j} = 0$ when $j \notin \mathcal{N}_i$. Let $\mathbf{e}(t) = (z_1(t), \ldots, z_n(t))^T - (z_\star, \ldots, z_\star)^T$ be the error vector at iteration t. The number of iterations t necessary for reaching $||\mathbf{e}(t)||_2/||\mathbf{e}(0)||_2 < \varepsilon$ ranges between a single iteration for complete graphs, and order $n^2 \log(\varepsilon^{-1})$ iterations for networks with lower connectivity like string and circular graphs [1, Theorems 1.79 and 1.80].

© The Author(s) 2015
R. Aragues et al., *Parallel and Distributed Map Merging and Localization*,
SpringerBriefs in Computer Science, DOI 10.1007/978-3-319-25886-7

A common choice for the matrix $W \in \mathbb{R}^{n \times n}$ is given by the Metropolis weights given by [3],

$$
W_{i,j} = \begin{cases} \frac{1}{1 + \max\{|\mathcal{N}_i|, |\mathcal{N}_j|\}} & \text{if } j \in \mathcal{N}_i, j \neq i, \\ 0 & \text{if } j \notin \mathcal{N}_i, j \neq i, \\ 1 - \sum_{j \in \mathcal{N}_i} W_{i,j}, & \text{if } j = i, \end{cases} \tag{A.3}
$$

for $i, j \in \mathcal{V}, j \neq i$, where $|\mathcal{N}_i|, |\mathcal{N}_j|$ are the number of neighbors of robots i, j. Note that each robot can compute the weights that affect its evolution using only local information. The algorithm (A.1) using the Metropolis weights W converges to the average of the inputs.

References

1. F. Bullo, J. Cortes, S. Martinez, *Distributed Control of Robotic Networks*. Applied mathematics series. (Princeton University Press, Princeton, 2009). http://coordinationbook.info
2. W. Ren, R. W. Beard. *Distributed Consensus in Multi-vehicle Cooperative Control*. Communications and control engineering. (Springer, London, 2008)
3. L. Xiao, S. Boyd, S. Lall. A scheme for robust distributed sensor fusion based on average consensus. In *Symposium on Information Processing of Sensor Networks (IPSN)*. Los Angeles, (2005), pp. 63–70

Appendix B
Auxiliary Results for Distributed Localization

Development of the Expressions of the Planar Localization Algorithm in Sect. 3.3.1

During the first phase, $\tilde{\theta}^a_{\mathcal{Y}a}$ and its covariance $\Sigma_{\tilde{\theta}^a_{\mathcal{Y}a}}$ are

$$\tilde{\theta}^a_{\mathcal{Y}a} = \Sigma_{\tilde{\theta}^a_{\mathcal{Y}a}} \mathcal{A}^a \Sigma_{\mathbf{z}_\theta}^{-1} \mathbf{z}_\theta, \qquad \Sigma_{\tilde{\theta}^a_{\mathcal{Y}a}} = (\mathcal{A}^a \Sigma_{\mathbf{z}_\theta}^{-1}(\mathcal{A}^a)^T)^{-1}. \qquad (B.1)$$

In the second phase, the updated measurements $\mathbf{w}$ and a first-order propagation of the uncertainty $\Sigma_{\mathbf{w}}$ are

$$\mathbf{w} = \begin{bmatrix} \tilde{\mathbf{z}}_{xy} \\ \tilde{\theta}^a_{\mathcal{Y}a} \end{bmatrix} = \begin{bmatrix} \tilde{R}\mathbf{z}_{xy} \\ \tilde{\theta}^a_{\mathcal{Y}a} \end{bmatrix}, \qquad \Sigma_{\mathbf{w}} = \begin{bmatrix} \tilde{R}\Sigma_{\mathbf{z}_{xy}}\tilde{R}^T + J\Sigma_{\tilde{\theta}^a_{\mathcal{Y}a}}J^T & J\Sigma_{\tilde{\theta}^a_{\mathcal{Y}a}} \\ \Sigma_{\tilde{\theta}^a_{\mathcal{Y}a}}J^T & \Sigma_{\tilde{\theta}^a_{\mathcal{Y}a}} \end{bmatrix}^T. \qquad (B.2)$$

The estimates in the third phase are the solution of the linear system

$$\hat{\mathbf{p}}^a_{\mathcal{Y}a} = \begin{bmatrix} \hat{\mathbf{x}}^a_{\mathcal{Y}a} \\ \hat{\theta}^a_{\mathcal{Y}a} \end{bmatrix} = (B\Sigma_{\mathbf{w}}^{-1}B^T)^{-1}B\Sigma_{\mathbf{w}}^{-1}\mathbf{w}. \qquad (B.3)$$

To write in explicit form $\hat{\mathbf{x}}^a_{\mathcal{Y}a}$ and $\hat{\theta}^a_{\mathcal{Y}a}$ we first compute the information matrix $\Upsilon_{\mathbf{w}} = \Sigma_{\mathbf{w}}^{-1}$,

$$\Upsilon_{\mathbf{w}} = \begin{bmatrix} \Upsilon_{\tilde{\mathbf{z}}_{xy}} & -\Upsilon_{\tilde{\mathbf{z}}_{xy}}J \\ -J^T\Upsilon_{\tilde{\mathbf{z}}_{xy}} & \Sigma_{\tilde{\theta}^a_{\mathcal{Y}a}}^{-1} + J^T\Upsilon_{\tilde{\mathbf{z}}_{xy}}J \end{bmatrix}, \qquad (B.4)$$

© The Author(s) 2015
R. Aragues et al., *Parallel and Distributed Map Merging and Localization*,
SpringerBriefs in Computer Science, DOI 10.1007/978-3-319-25886-7

where $\Upsilon_{\tilde{\mathbf{z}}_{xy}}$ is as in Eq. (3.11), $\Upsilon_{\tilde{\mathbf{z}}_{xy}} = (\tilde{R}\Sigma_{\mathbf{z}_{xy}}\tilde{R}^T)^{-1}$, and where we have used the following blockwise inversion relations $\begin{bmatrix} A & B \\ C & D \end{bmatrix}^{-1} = \begin{bmatrix} E & F \\ G & H \end{bmatrix}$, with

$$
\begin{aligned}
E &= A^{-1} + A^{-1}B\left(D - CA^{-1}B\right)^{-1}CA^{-1} = \left(A - BD^{-1}C\right)^{-1}, \\
F &= -A^{-1}B\left(D - CA^{-1}B\right)^{-1} = -\left(A - BD^{-1}C\right)^{-1}BD^{-1}, \\
G &= -\left(D - CA^{-1}B\right)^{-1}CA^{-1} = -D^{-1}C\left(A - BD^{-1}C\right)^{-1}, \\
H &= \left(D - CA^{-1}B\right)^{-1} = D^{-1} + D^{-1}C\left(A - BD^{-1}C\right)^{-1}BD^{-1}.
\end{aligned}
\tag{B.5}
$$

The information matrix $\Upsilon_{\hat{\mathbf{p}}_{\mathscr{Y}a}} = (B\Sigma_{\mathbf{w}}^{-1}B^T)$ and its inverse $\Sigma_{\hat{\mathbf{p}}_{\mathscr{Y}a}}$ are

$$
\Upsilon_{\hat{\mathbf{p}}_{\mathscr{Y}a}} = \begin{bmatrix} (\mathscr{A}^a \otimes \mathbf{I}_2)\Upsilon_{\tilde{\mathbf{z}}_{xy}}(\mathscr{A}^a \otimes \mathbf{I}_2)^T & -(\mathscr{A}^a \otimes \mathbf{I}_2)\Upsilon_{\tilde{\mathbf{z}}_{xy}}J \\ -J^T\Upsilon_{\tilde{\mathbf{z}}_{xy}}(\mathscr{A}^a \otimes \mathbf{I}_2)^T & \Sigma_{\tilde{\theta}_{\mathscr{Y}a}}^{-1} + J^T\Upsilon_{\tilde{\mathbf{z}}_{xy}}J \end{bmatrix},
$$

$$
\Sigma_{\hat{\mathbf{p}}_{\mathscr{Y}a}} = \begin{bmatrix} \Sigma_{\hat{\mathbf{x}}} & \Sigma_{\hat{\mathbf{x}},\hat{\theta}} \\ \Sigma_{\hat{\mathbf{x}},\hat{\theta}}^T & \Sigma_{\hat{\theta}} \end{bmatrix}, \text{ with}
\tag{B.6}
$$

$$
\begin{aligned}
\Sigma_{\hat{\theta}} &= ((\Sigma_{\tilde{\theta}_{\mathscr{Y}a}})^{-1} - J^T\Upsilon_{\tilde{\mathbf{z}}_{xy}}EJ)^{-1}, \\
\Sigma_{\hat{\mathbf{x}}} &= L^{-1} + L^{-1}(\mathscr{A}^a \otimes \mathbf{I}_2)\Upsilon_{\tilde{\mathbf{z}}_{xy}}J\Sigma_{\hat{\theta}}J^T\Upsilon_{\tilde{\mathbf{z}}_{xy}}(\mathscr{A}^a \otimes \mathbf{I}_2)^TL^{-1}, \\
\Sigma_{\hat{\mathbf{x}},\hat{\theta}} &= L^{-1}(\mathscr{A}^a \otimes \mathbf{I}_2)\Upsilon_{\tilde{\mathbf{z}}_{xy}}J\Sigma_{\hat{\theta}} \\
E &= (\mathscr{A}^a \otimes \mathbf{I}_2)^TL^{-1}(\mathscr{A}^a \otimes \mathbf{I}_2)\Upsilon_{\tilde{\mathbf{z}}_{xy}} - \mathbf{I}, \\
L &= (\mathscr{A}^a \otimes \mathbf{I}_2)\Upsilon_{\tilde{\mathbf{z}}_{xy}}(\mathscr{A}^a \otimes \mathbf{I}_2)^T.
\end{aligned}
\tag{B.7}
$$

Z_n^p and M_n^p Matrices Defined by Blocks in the Proof of Theorem 6

In [1], a classification of matrices defined by blocks and a study of their properties is given. Here we show a brief summary of some of these properties. We use the notation $A = [A_{ij}]$ for a real matrix $A \in \mathbb{R}^{np \times np}$ defined by blocks, where each block A_{ij} is a $p \times p$ matrix, for all $i, j \in \{1, \ldots, n\}$.

Definition 6 ([1]) Matrix A is of class Z_n^p if A_{ij} is symmetric for all $i, j \in \{1, \ldots, n\}$ and $A_{ij} \preceq \mathbf{0}$ for all $i, j \in \{1, \ldots, n\}, j \neq i$. In addition, it is of class $\hat{Z}_n^p$ if $A \in Z_n^p$ and $A_{ii} \succ \mathbf{0}$ for all $i \in \{1, \ldots, n\}$. Matrix A is of class M_n^p if $A \in \hat{Z}_n^p$, and there exist positive scalars $u_1, \ldots, u_n > 0$ such that

$$
\sum_{j=1}^{n} u_j A_{ij} \succ \mathbf{0} \text{ for all } i \in \{1, \ldots, n\}.
$$

Lemma 4 *([1, Lemma 3.8]) Let $A \in Z_n^p$ and assume that $\forall \, \mathcal{J} \subset \{1, \ldots, n\}$, there exists $i \in \mathcal{J}$ such that $\sum_{j \in \mathcal{J}} A_{ij} \succ 0$. Then, there exists a permutation π such that $\sum_{j \geq i} A_{\pi(i),\pi(j)} \succ 0$, for all $i \in \{1, \ldots, n\}$.*

Theorem 9 *([1, Theorem 3.11]) Let $A \in Z_n^p$, let $u_1, \ldots, u_n > 0$ and let*

$$\sum_{j=1}^{n} A_{ij} u_j \succeq 0, \text{ for all } i \in \{1, \ldots, n\}. \tag{B.8}$$

Assume that there exists a permutation π of $\{1, \ldots, n\}$ such that

$$\sum_{j \geq i} A_{\pi(i),\pi(j)} u_{\pi j} \succ 0, \text{ for all } i \in \{1, \ldots, n\}. \tag{B.9}$$

Then, $A \in M_n^p$.

Theorem 10 *([1, Theorem 4.7]) Let*

$$A + A^T \in M_n^p, \qquad D = \text{blkDiag}(A_{11}, \ldots, A_{nn}), \qquad \text{and} \quad A = D - N.$$

Then $\rho\left(D^{-1}N\right) < 1$.

Reference

1. L. Elsner, V. Mehrmann, Convergence of block iterative methods for linear systems arising in the numerical solution of Euler equations. Numerische Mathematik **59**(1):541–559 (1991)

Index

© The Author(s) 2015
R. Aragues et al., *Parallel and Distributed Map Merging and Localization*,
SpringerBriefs in Computer Science, DOI 10.1007/978-3-319-25886-7